t.f.h.

nudibranchs

dr. t.e. thompson

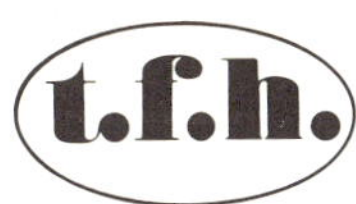

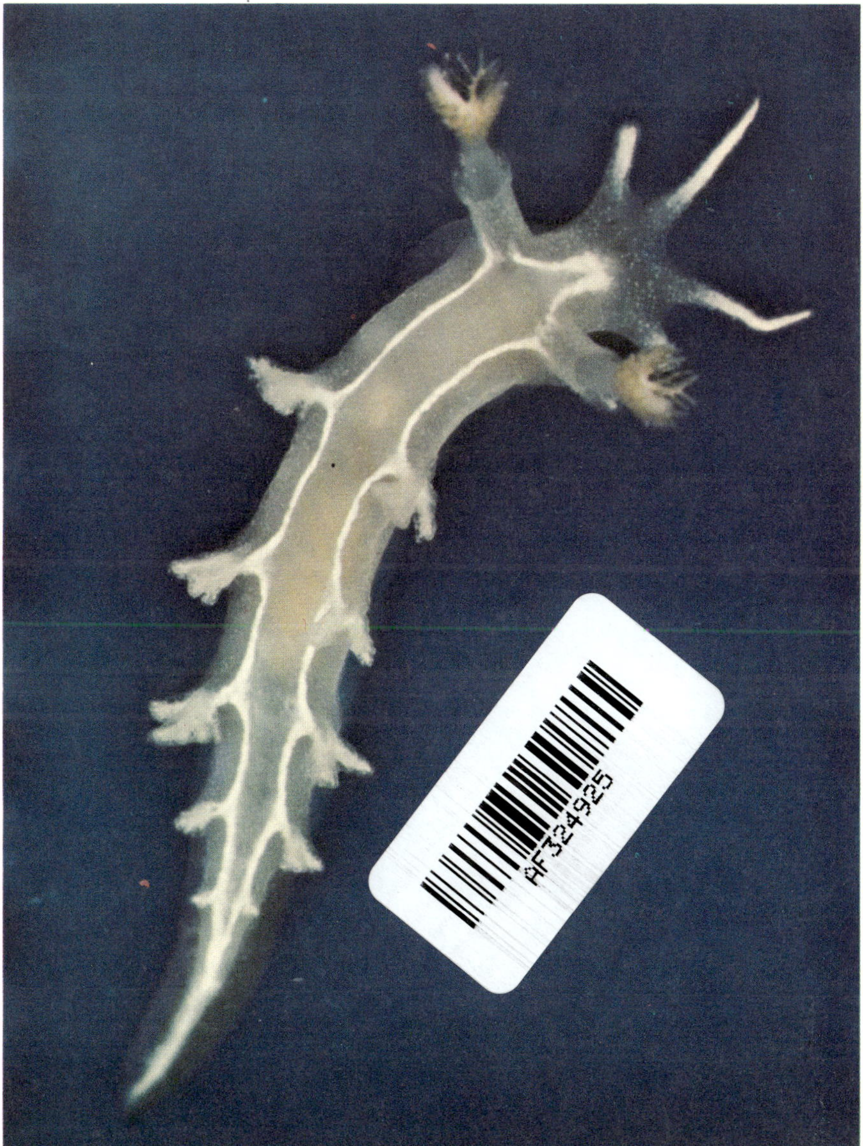

© 1976 by T.F.H. PUBLICATIONS, INC., LTD.

ISBN 0-87666-459-1

Distributed in the U.S.A. by T.F.H. Publications, Inc., 211 West Sylvania Avenue, P.O. Box 27, Neptune City, N.J. 07753; in England by T.F.H. (Gt. Britain) Ltd., 13 Nutley Lane, Reigate, Surrey; in Canada to the book store and library trade by Clarke, Irwin & Company, Clarwin House, 791 St. Clair Avenue West, Toronto 10, Ontario; in Canada to the pet trade by Rolf C. Hagen Ltd., 3225 Sartelon Street, Montreal 382, Quebec; in Southeast Asia by Y.W. Ong, 9 Lorong 36 Geylang, Singapore 14; in Australia and the south Pacific by Pet Imports Pty. Ltd., P.O. Box 149, Brookvale 2100, N.S.W., Australia. Published by T.F.H. Publications, Inc. Ltd., The British Crown Colony of Hong Kong.

Contents

Nudibranchs are all carnivores. The North Atlantic *Tritonia hombergi* is shown here attacking the soft coral *Alcyonium digitatum*; it can bite out a dime-sized chunk.

Above: Doto pinnatifida exhibits swollen cerata on each side of the back. *Doto* is a dendronotacean genus found in all oceans. *Below:* Many dendronotaceans are intricately camouflaged. Here *Tritonia odhneri* feeds on the gorgonian *Eunicella verrucosa*.

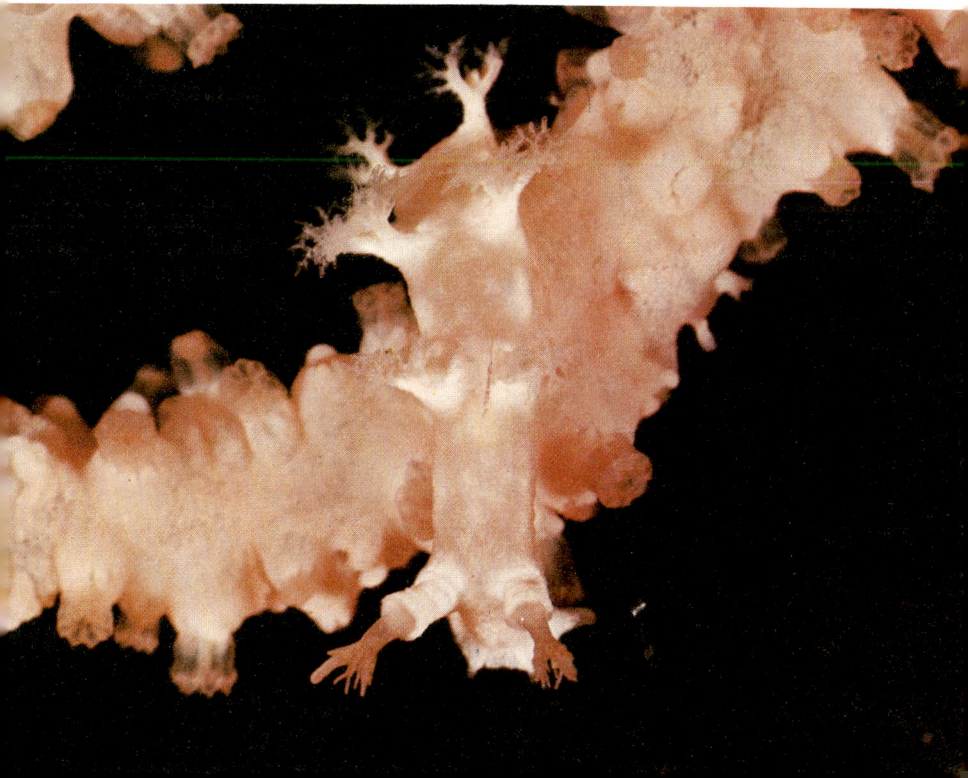

Introduction

The Nudibranchia are an order of opisthobranch gastropod mollusks—the true sea-slugs ; they are all macroscopic (the commonest species ranging from 2 to 200 mm in length), marine (although some nudibranchs can withstand a wide range of salinity), and hermaphrodite (with both sets of sexual organs functioning in the same body simultaneously). World-wide there are approximately 2500 species; nearly all of these live *on* the sea bottom, not *within* it or *above* it. Some of the rarest and most interesting nudibranchs defy these rules and burrow into soft marine substrates or the bodies of their prey or, alternatively, swim like graceful dancers in the surface waters of the ocean. Nudibranchs attracted the devoted attention of naturalists in the last century and now, in the nineteen-seventies, have intrigued and captured the interest of the growing body of undersea aqualung enthusiasts.

Nudibranchs are to the phylum Mollusca what the orchids are to the angiosperms (flowering plants) or the butterflies are to the arthropods. Although their living species cannot rival the prosobranch gastropods in terms of numbers, they show more variety in their behavior, diets, defensive adaptations, and in the flamboyance of their body ornamentation. In fact they are among the most visually exciting invertebrate animals. They range in size from the tiny *Pseudovermis*, which is so minute that it can move unharmed between the sand grains that make up a beach, to the huge *Tochuina tetraquetra* of the Pacific Northwest of America, which can weigh in air as much as 1½ kilograms. Some may be found only by diving, digging, or searching between boulders or coral heads. Others, like some of the tropical chromodorid nudibranchs, often appear bold and self-advertising.

Although *Tochuina tetraquetra* was in the past taken for human consumption in the Aleutian Islands and the dried bodies of *Aplysia* (not a true sea-slug but fairly closely related) are used medicinally by physicians in coastal regions of China, it is certainly well established that handling some living opisthobranchs can lead rapidly to mild but unpleasant inflammation and pain in the human skin. The British nudibranch *Tritonia hombergi*, for instance, can discharge a brownish secretion that can blister the human hand. A recent report established that the planktonic aeolidacean nudibranch *Glaucus atlanticus*, which often feeds upon the venomous siphonophore *Physalia* (the Portuguese Man-of-War), utilizes for its own defense the nematocysts (stinging cells) of this formidable prey. Stings by *Glaucus* resembling those inflicted by *Physalia* have been reliably reported in Australia by bathers who misguidedly handled these pretty blue mollusks.

To the comparative zoologist the nudibranchs represent an interesting evolutionary offshoot of the Mollusca in which the precarious balance between the hard and the soft parts has become heavily tilted in favor of the latter. Like other mollusks, they are coelomate invertebrates, although the true coelom is reduced in volume, remaining detectable only in the lumina of the renal, pericardial, and gonadial sacs. The blood-filled haemocoelic perivisceral spaces of a nudibranch mollusk are not coelomic in character, but are derived from embryonic blastocoelic cavities. It would be incorrect, however, to think that the arterial systems of these animals are flaccid and aimless, although it is true that the venous system can be voluminous and relatively slow and inefficient. The central nervous system of the remote ancestor of the nudibranchs doubtless consisted of a circumesophageal ring of ganglia connected by commissures and connectives with each other and with certain visceral ganglionic masses. In the present-day nudibranch mollusks, evolutionary trends can be detected toward cephalization (shortening of connectives, resulting in the aggregation of the ganglia around the foregut) and toward cerebralization (blending of individual ganglia).

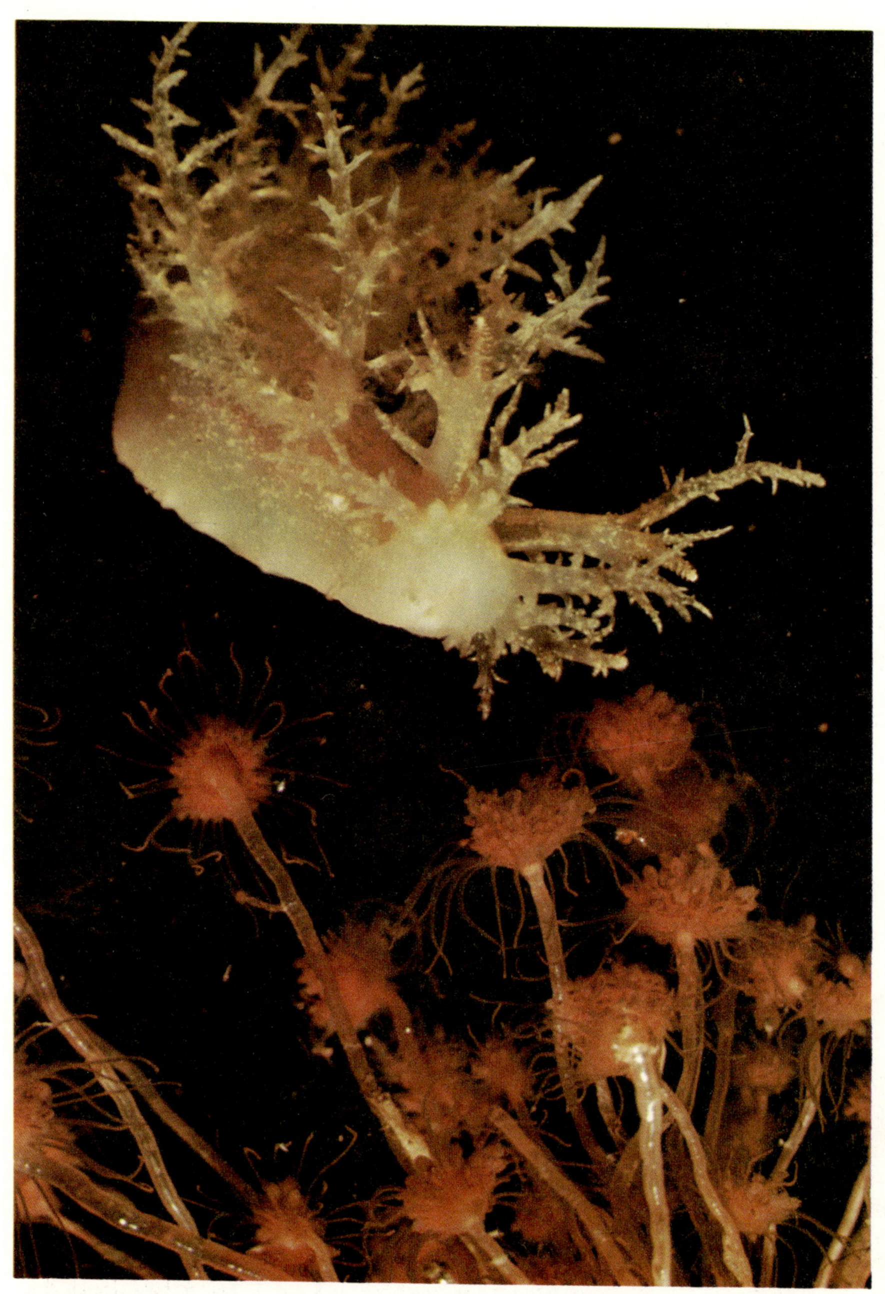

Dendronotus frondosus is typical of the dendronotaceans in feeding on coelenterates. Here it is preparing to crop the hydroid *Tubularia*.

Above: The rare and delicate *Lomanotus marmoratus* shows typical dendronotacean characters of slender elongated body, tree-like gill processes, and basal sheaths on the rhinophoral tentacles. *Below: Tethys fimbria* is an exceptional dendronotacean which feeds upon small crustaceans captured with the dilated head lobes.

Evolutionary History

Within the gastropod mollusks the systematic zoologist recognizes two major categories. The first is the Streptoneura (or Prosobranchia), which usually have a heavy, coiled external shell, live predominantly in the sea, and exhibit internally the effects of visceral twisting or torsion. The second category is the Euthyneura; they are usually more lightly shelled and the operculum, which was used to close the 'door' of the shell in their remote ancestors, has almost invariably now been lost. Euthyneura is a technical term implying that the nervous system is untwisted, and indeed most modern euthyneurans exhibit only slight traces of the ancestral torsional twisting. Because this euthyneuran category includes the nudibranchs, it is necessary to look more closely at the evolutionary history of this great group, which comprises also the pulmonate slugs and snails of the land and fresh water.

In attempting to seek evidence about the evolutionary origins and early history of the Euthyneura, one is forced to admit that hard-and-fast distinctions between the early Opisthobranchia (the branch which led eventually to the Nudibranchia) and the early Pulmonata (the branch which led to most present-day land and fresh water slugs and snails) become difficult to define. Probably the earliest euthyneurans took their origin from some benthic, epifaunal, helically coiled, strong-shelled marine prosobranch gastropod stock mid-way through the Paleozoic. Conceivably this stock was close in structure and habits to many present-day rissoaceans, which are prosobranch mesogastropods, although not all authorities are agreed on this derivation and one major expert seeks the evolutionary origin of the Euthyneura among the early Archaeogastropoda. Some time later

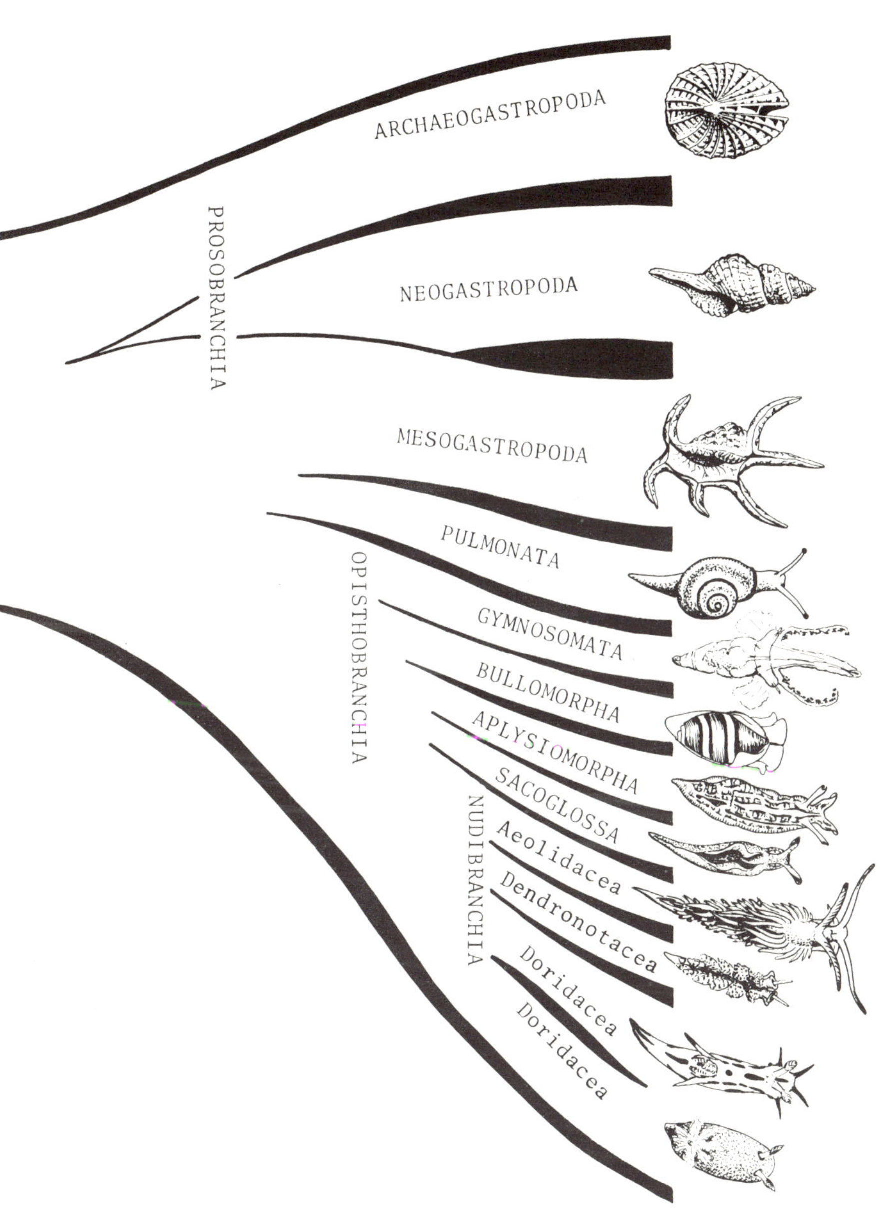

Evolutionary relationships of the Nudibranchia.

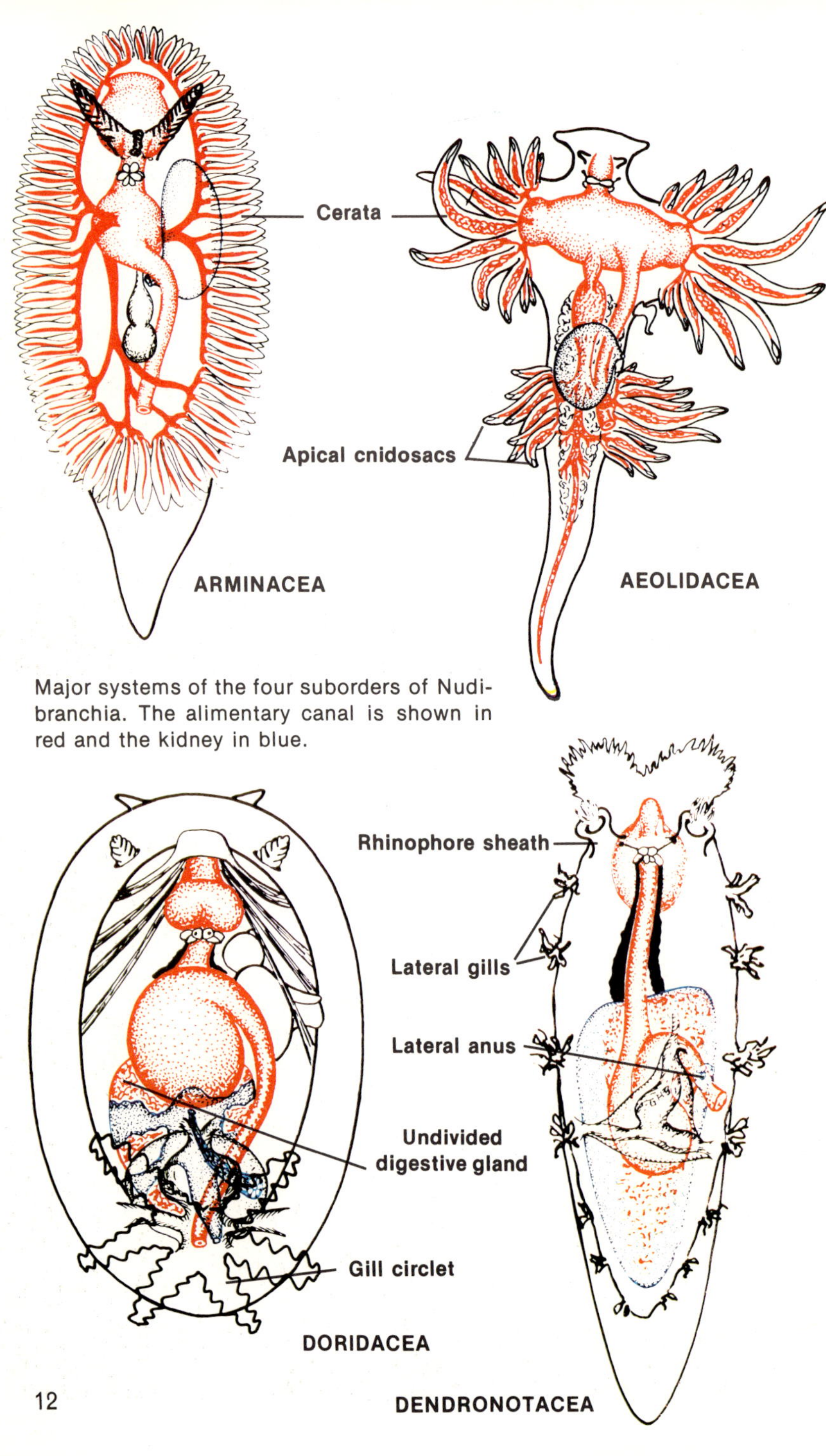

Major systems of the four suborders of Nudibranchia. The alimentary canal is shown in red and the kidney in blue.

Swimming in the dendronotacean *Dendronotus iris. Above:* Creeping. *Below:* Alarmed specimen swimming by threshing the body from side to side. In this species swimming is purely an escape reaction and ceases when the danger has passed.

the euthyneuran line split into two successful branches, one succeeding in the conquest of brackish and fresh water, leading on to the invasion of purely terrestrial habitats; this became the Pulmonata. The other line led to the exploitation of the marine infaunal (burrowing) environment and produced by the late Paleozoic the earliest Opisthobranchia. It is interesting to note that these two major lines of descent allowed the Gastropoda access for the first time to fresh water and the land in the first case, and to marine burrowing in the other. The ancestral prosobranch stocks had not succeeded in invading such habitats nor have their prosobranch descendants exhibited very much more success in these respects.

The two main lines of descent from the first euthyneurans have many features in common. In both lines the static, negative defensive mechanism represented by the operculum-shell soon came to be of reduced importance and was replaced by more dynamic biological and chemical methods of defense. The shells of euthyneurans are nearly always more frail than those of the prosobranchs, and the operculum is seldom present in pulmonates and exists in only a small number of the most primitive living opisthobranchs. The effects of gastropod torsion have been progressively nullified then abolished in both pulmonate and opisthobranch lines of descent, and the most advanced members of both groups show external and internal bilateral symmetry of all systems except the reproductive apparatus, which usually remains predominantly on the right side of the body in connection with the pairing posture suitable for reciprocal copulation in such hermaphrodite animals.

The earliest opisthobranchs probably all became burrowing infaunal animals. Certainly the structurally primitive living opisthobranchs (*Acteon, Pupa, Scaphander*) are burrowers. In these early forms the body became enlarged and anteriorly spatulate; the head became a flattened spade-like cephalic shield. The head tentacles, originally placed on top, were lost or became shifted to the sides or rear of the head. The mantle cavity, formerly placed anteriorly, owing to the torsion shown by the ancestral forms became more posteriorly situated and open to avoid clogging and fouling during the movements of burrowing. The shell became more streamlined in shape and elongated antero-posteriorly, so as to permit subterranean locomotion, and the primitive coiling of the visceral hump was lost as the digestive gland, stomach, and ovotestis came to be accommodated as much within the mesopodium as in the increasingly ineffectual shell. Great lateral (parapodial) lobes of the foot gradually developed. These, together with newly evolving posteriorly directed flattened lobes of the cephalic shield, helped to enhance streamlining and to exclude foreign particles from the shell and mantle cavities.

From the early burrowing opisthobranchs with their growing emphasis on chemical and biological defense and concomitant reduction in the importance of the shell, together with their decisive and probably irreversible evolutionary loss of the operculum, arose a number of quite distinct lines of descent, the modern representatives of each of which constituted a group of ordinal rank. Just as the initial drive which was responsible for the early evolutionary experiments with the burrowing life described above was probably dietary and directed toward the exploitation of the rich annelid, protozoan, and bivalve molluscan infaunal (burrowing) food potential, so each of the orders of living opisthobranchs shows groups of species, genera, and families which have taken up new and quite different, often rather restricted, dietary specializations.

In the largest and most varied order of opisthobranchs, the Nudibranchia, there is some evidence of polyphyletic descent from a number of long-extinct opisthobranch stocks. In all those lines of nudibranchs which persist to the present day, the shell and operculum have been discarded in the adult form. In many of them the body has quite independently become dorsally papillate. In at least two suborders these dorsal papillae have acquired the power to nurture nematocysts derived from cnidarian prey so as to use them for the nudibranch's own defense. In other cases such cerata have independently become penetrated by lobules of the adult digestive gland, or they may contain virulent defensive glands or prickly bundles of calcareous defensive spicules. So

Above: Casella atromarginata, a doridacean, is very tough, the skin feeling like wet leather. It feeds on spiculose sponges. The genus *Casella* is widespread in the Indo-Pacific. *Below:* The flattened *Armina californica* burrows in sand to hunt its prey, the sea pansy *Renilla.*

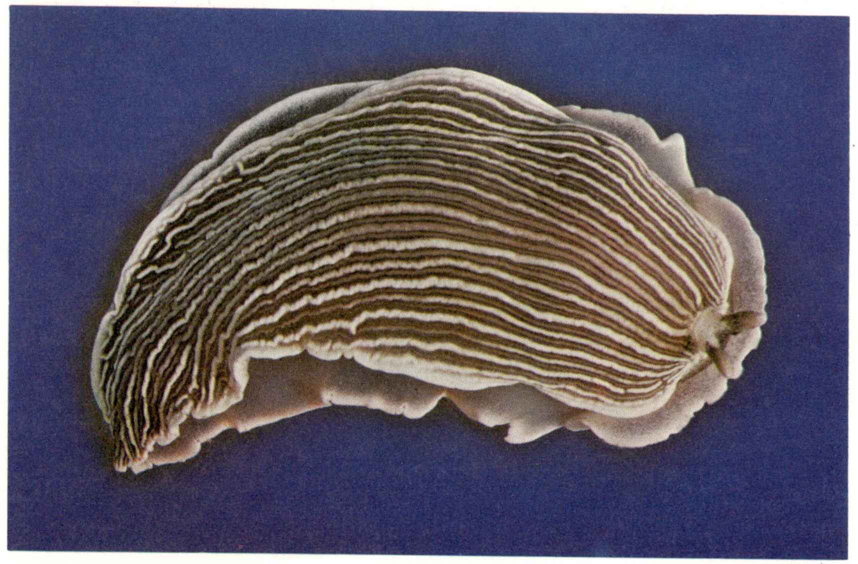

Both *Armina* and *Dirona albolineata* (above) are arminacean nudibranchs, but *Dirona* has rows of dorsal papillae which cause it to resemble aeolidaceans. The smaller specimen shows the curled white egg sacs of the internal parasitic copepod *Splanchnotrophus*; nudibranchs attacked by this copepod may live for weeks, but they usually fail to reach sexual maturity.

pervasive are the tendrils of the converging lines of evolution that the precise interrelationships between the dendronotacean, arminacean, aeolidacean, and doridacean suborders may never be fully understood.

In the Nudibranchia can be seen the great advantage acquired by gastropods which have exploited to the full the evolutionary loss of the constraints of passive mechanical defense represented by the shell and operculum and have replaced these by active dynamic defensive adaptations.

Nudibranchs are found in all the world's oceans and major seas; they feed on every kind of epifaunal animal material, as well as playing a significant part in some planktonic communities. Some nudibranchs, such as the aeolidacean *Cerberilla*, actually burrow into soft marine deposits in search of infaunal sea anemones. They owe their success to their unique maintenance, during their evolutionary history, of the delicate balance between the shell and skin as defensive attributes. They have remained viable during the long transition from the primitive condition, much as in modern *Acteon*, to the present-day carnivorous nudibranch species. These have agile, soft bodies, perfectly adapted chemosensory tentacles (rhinophores) and buccal apparatus, and virtual immunity to attack by fish and many other predators, while the searching larvae have an extraordinary ability to recognize and then metamorphose upon the adult diet.

Collection, Preservation and Systematics

The larger nudibranchs are often easy to find between tide marks on rocky shores, sometimes in shallow water or in pools. The smaller species present considerable problems and can often be found only after patient and laborious search through dredged hydroids, sponges, and bryozoans. Aqualung divers are privileged to see the sublittoral species in their natural habitats.

The best localities are often the most hazardous and inaccessible. In temperate areas exposed headlands or rocky islands in areas of rapid currents with water of high salinity are usually the most favorable. In warmer waters coral reefs can be disappointing. This is because coral polyps have tentacles that are armored with nematocysts, and nudibranchs creeping 'barefoot' as they do are repelled by this. But coral rubble such as may commonly be found in a quiet lagoon in the lee of a coral reef is a very good place to hunt for nudibranchs.

Spawn coils often give a clue to the whereabouts of the adult nudibranchs, but many nudibranchs are conspicuous anyway and such detective work may be unnecessary.

Obviously no responsible naturalist will take more animals than are needed for his immediate purpose, and, in fact, great satisfaction may be obtained from simple ecological observations without the need for capturing and killing the subjects. With difficult individuals or obscure species, however, it is certainly not practical to attempt identification in the field and it is usually found that nudibranchs can be transported safely to the laboratory or the aquarium in clean sea water in a stout plastic bag (or in a vacuum flask if

Above: The doridacean *Limacia clavigera* has a pattern of brilliant yellows on white. *Below:* The delicate *Goniodoris nodosa* feeds on the soft parts of encrusting ascidians (sea squirts). It belongs to the doridacean family Goniodorididae.

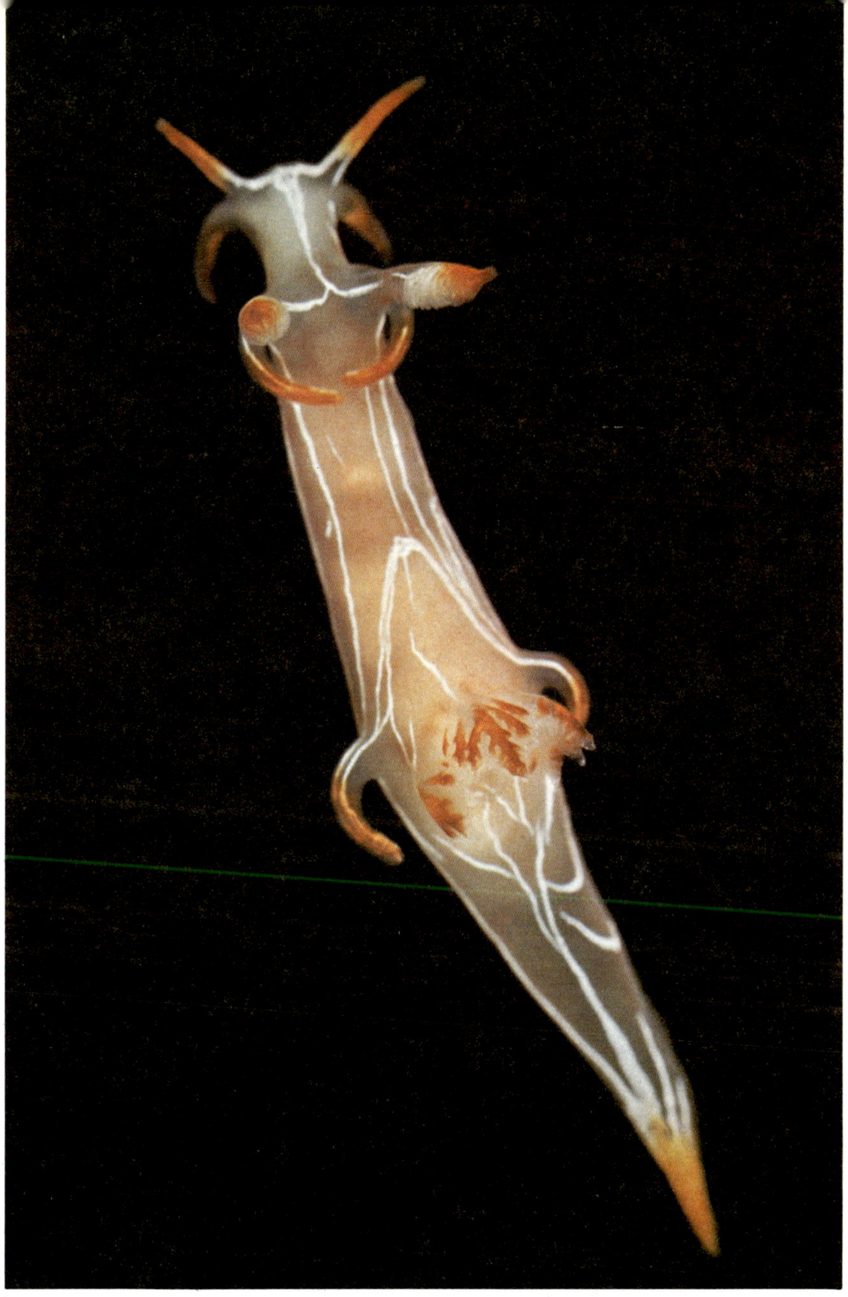

Warning colors which take the form of bright yellow-tipped glandular papillae are present in many species of doridacean nudibranchs. Is this a mere coincidence or is it perhaps the case that a mutual advantage can be gained by speedy "education" of potential predators to avoid such unpleasant morsels? Shown here is *Trapania lineata* from the Bay of Naples.

the climate is extreme). In the laboratory the specimens should be transferred to cool clean sea water and not over-crowded. Remember that some of the larger species are voracious when hungry and may attack more delicate specimens.

Observations from life are especially important in these soft and highly contractile animals. This is a challenging area for inquiry, because Japanese workers have recently succeeded in tagging individual marine slugs with plastic-coated wire and DYMO tape. Preservation should only follow careful noting of habitat, morphology, and colors. Narcotization may be effected with 7% aqueous magnesium chloride mixed in equal parts with sea water. The animals when relaxed can be preserved in 10% formalin and may later be transferred to 70% alcohol if this is desired.

A detailed discussion of the internal anatomy of the Nudibranchia is outside the scope of this book. The interested reader should consult the Ray Society Monograph by the author (*Biology of Opisthobranch Molluscs*, Volume I. 1976. British Museum, London). It is necessary to say something about the radula of the nudibranchs because this organ is different in every nudibranch species and provides the patient investigator with a check on identifications made from external features alone. Other anatomical features which must be understood before we address ourselves to the systematic zoology of nudibranchs are shown diagrammatically in the drawings. They show the principal features of typical representatives of the four nudibranch suborders Dendronotacea, Arminacea, Aeolidacea, and Doridacea.

The radula of a nudibranch is made of chitin similar to that found in the exoskeletons of insects and crustaceans. Nudibranchs have specifically distinct radulae because each species has its own feeding specialty and teeth adapted for dealing with its prey. This scanning electron micrograph shows part of the radula of *Casella atromarginata*, a sponge feeder.

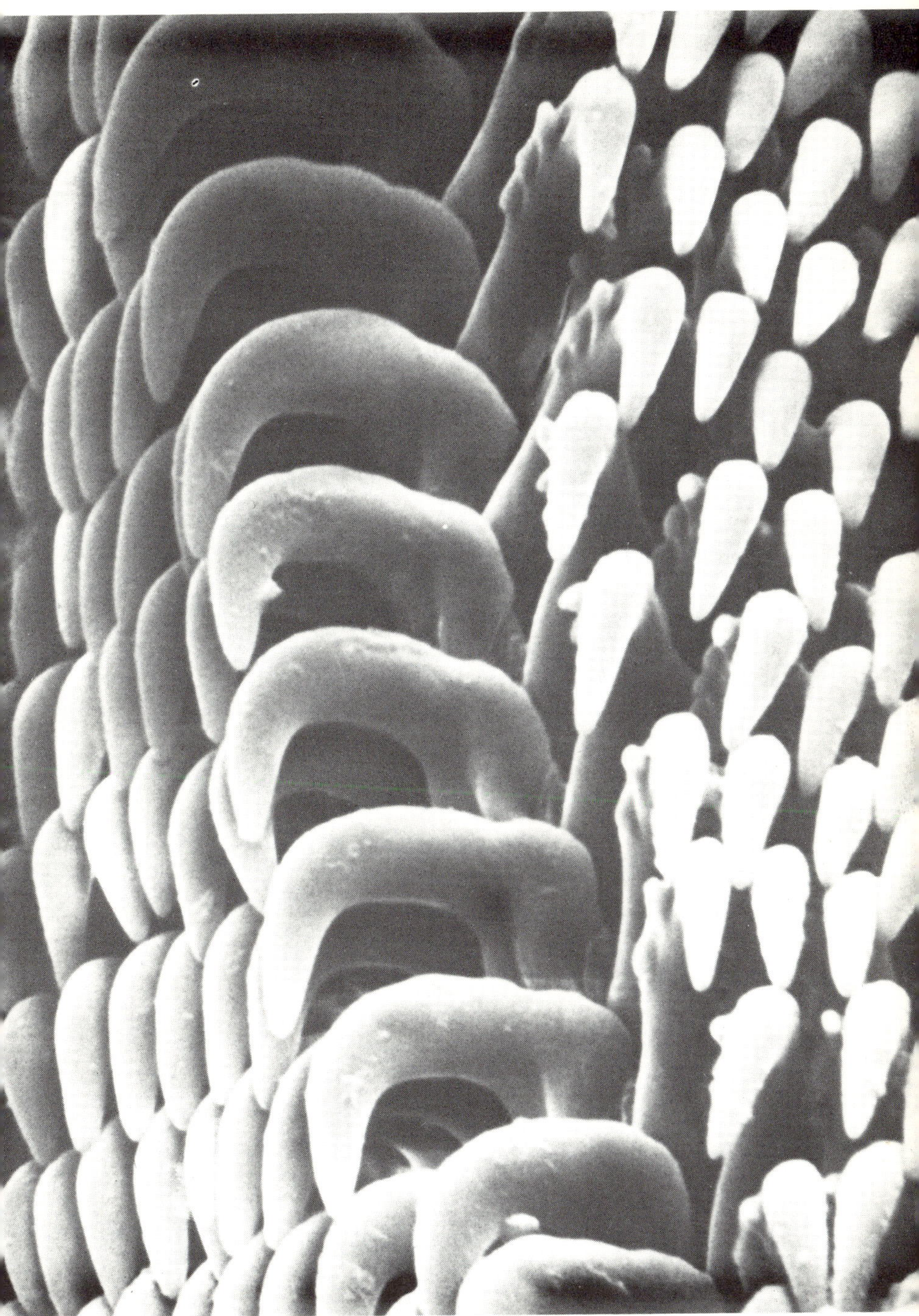

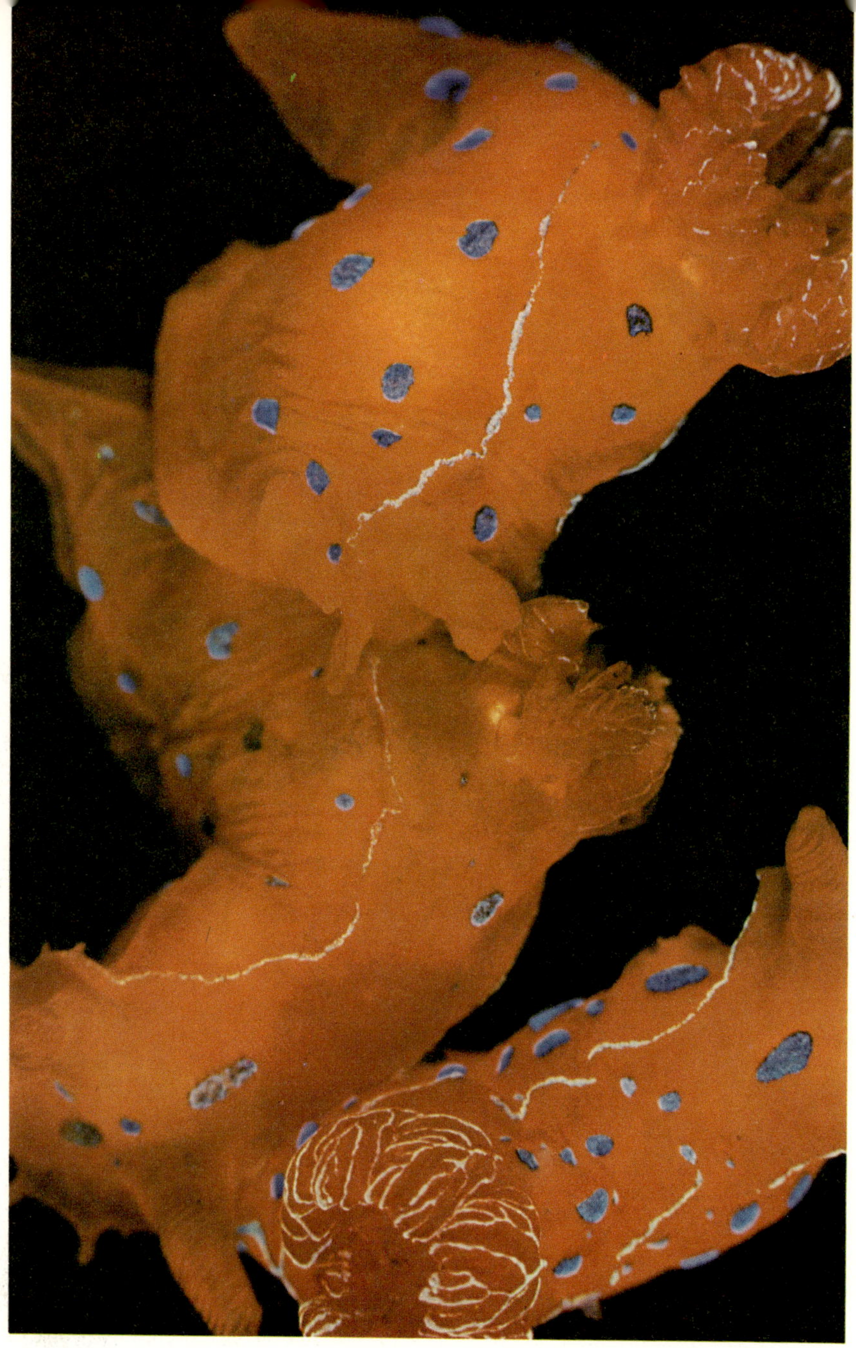

The brilliant blue spots on orange of *Greilada elegans* function to warn potential predators of its distasteful properties (repellent glands under the gaudy skin).

Ceratosoma cornigerum holds up a fleshy lure to divert attention from the gills. The lure is filled with defensive glands that discharge if bitten by a fish.

The radula of a nudibranch is readily seen with the un-aided eye in a larger nudibranch if a cut is made behind the mouth, but the radulae of many of the smaller species are microscopic and very difficult to dissect out of the body. The best method to use for the isolation and preparation for microscopy of such radulae is to dissect out the whole buccal bulb (the muscular organ inside the mouth which houses and manipulates the radula itself) from the spirit or formalin pre-served specimen. The buccal bulb is then warmed in 10% caustic soda or potash in order to disperse the flesh; this may take minutes or even hours but may be hastened by heating (NOT boiling) in a test tube. Hot caustic fluids are extremely

Radular teeth of the doridacean *Ceratosoma cornigerum*; its diet is not known.

dangerous to the human skin and destructive to human clothing, so every care should be taken. When the radula is free from tissue it can be transferred to a wash of 70% alcohol before mounting on a glass microscope slide. One drop of a mountant such as polyvinyl lactophenol and a thin cover glass complete the preparation. Of course, such mountants must be carefully dried or oven-baked before they can be regarded as permanent. A new student's initial efforts at radula preparation are often clumsy, special difficulty being experienced at the stage when the radula is laid on the microscope slide in an orientation that is easily interpretable. As in so many techniques which at first seem impossibly difficult, practice may make perfect. If you are short of nudibranchs to practice with, then use ordinary wild garden or farm snails—they have perfectly good radulae, too.

The photographs show various features of nudibranch radular morphology displayed with the aid of a scanning electron microscope. Such electronic sophistication is by no means essential, and much can be seen in a good lactophenol preparation using a moderately good optical microscope. Points to note in studying nudibranch radulae are the number of rows of teeth, the number of teeth per row, the presence or absence of a median tooth in each row, and of course the actual shapes and sizes of the individual teeth. A glance through the photographs will show the reader something of the range of form that may be encountered.

Much remains to be discovered about radulae; we are still uncertain, for instance, about the significance of the various tooth patterns. We can see fairly well how the broad radulae of some of the doridacean nudibranchs are adapted to scraping up the two-dimensional sponge encrustations that many of them feed upon. But why are the radulae of those nudibranchs that feed upon sea anemones and other coelenterates usually so narrow? In such discussions we must not forget that many nudibranchs possess stout chitinous jaws in addition to the radula in the buccal bulb. The jaws probably always serve to cut off slices from a three-dimensional food organism, the radula then acting to transfer these slices into the esophagus on its way to the stomach.

It is often said that the Atlantic nudibranchs are less brilliant than their Indo-Pacific counterparts, but many Caribbean species are just as vividly patterned as Indo-Pacific species. Typical of these beautiful Caribbean nudibranchs is the chromodorid *Chromodoris neona,* which ranges north to Florida.

Other Caribbean chromodorids, such as this *Hypselodoris bayeri,* also display beautiful and varied patterns.

Classification

Suborder I. DENDRONOTACEA

This suborder contains ten families of nudibranchs, all possessing a mid-lateral anal papilla and distinct, often elaborate sheaths for the rhinophoral tentacles to be pulled into after an alarm. They feed upon coelenterates as varied as medusae (in the case of the Mediterranean *Phylliroe buceph-*

Radula of the dendronotacean *Tritonia hombergi,* which feeds on alcyonarians.

ala), hydroids and anemones (in European and North American species of *Lomanotus, Hancockia, Dendronotus,* and *Doto*), or soft alcyonarian and gorgonian corals (preferred by the world-wide species of *Tritonia* and *Marionia*). Horny jaws may or may not be present, and the radula shows a great deal of variation in the different families, from broad and multiseriate (having many rows) (Tritoniidae) to uniseriate (*Doto*), resembling the radula of some of the aeolidaceans. The body has a dorso-lateral ridge on each side of the body, frequently bearing arborescent processes which function as gills. Sometimes these gills contain tributaries of the alimentary canal, and a dendronotacean such as *Doto* can be mistaken for a true aeolidacean by the unwary student (the presence of rhinophore sheaths in *Doto* is decisive evidence for its retention in the Dendronotacea). Finally, mention must be made here of *Tethys* and *Melibe*, which occur on American coasts and lack both jaws and radula. They feed by casting about in muddy eel-grass and other rich areas for small crustaceans which are captured with the aid of a dilated fimbriated buccal hood and swallowed whole.

FAMILY	REPRESENTATIVE GENERA
Tritoniidae	*Tritonia, Marionia, Tochuina*
Marianinidae (= Aranucidae)	*Marianina*
Lomanotidae	*Lomanotus*
Dendronotidae	*Dendronotus*
Hancockiidae	*Hancockia*
Bornellidae	*Bornella*
Dotoidae	*Doto*
Tethyidae (= Fimbriidae)	*Tethys, Melibe*
Scyllaeidae	*Scyllaea*
Phylliroidae	*Phylliroe, Cephalopyge*

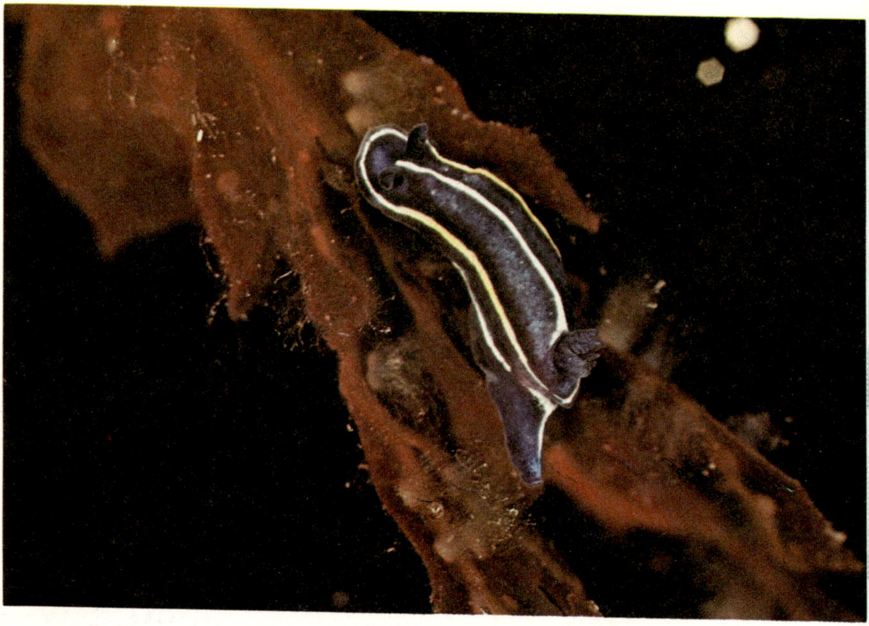

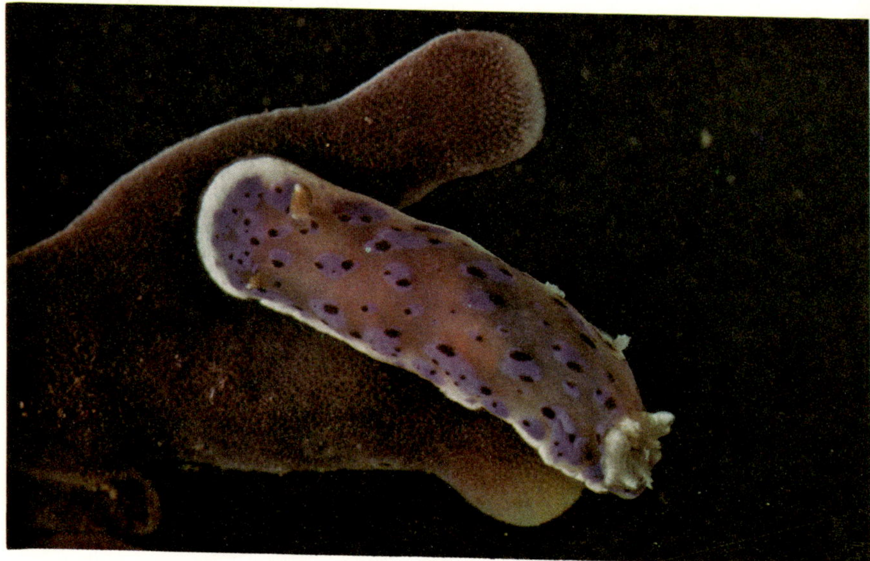

Above: Although chromodorids are usually found in warm seas, *Hypselo-doris tricolor* is common in the relatively cool Mediterranean Sea. It feeds on small encrusting sponges. *Below:* The Australian *Chromodoris loringi* is ecologically similar to *Hypselodoris tricolor,* also feeding on sponges.

Some doridaceans, such as these *Diaulula sandiegensis,* are not brightly colored. This species is inconspicuous when feeding on its usual prey, sponges.

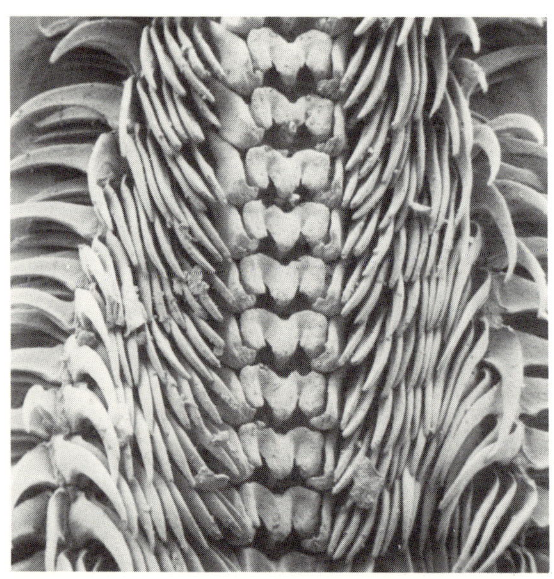

Dendronotacean radulae. *Above: Tritonia lineata*, feeds on alcyonarians. *Below left: Lomanotus marmoratus*, feeds on hydroids. *Below right: Dendronotus frondosus*, feeds on hydroids.

Suborder II. DORIDACEA

This is the largest nudibranch suborder, and in it are placed about 25 families. Most of these families are distributed throughout the world's seas, but some of them (Platydorididae, Asteronotidae, Baptodorididae, Actinocyclidae, Bathydorididae, and Doridoxidae) are absent or rare in cooler Northern Hemisphere waters, while others are quite decidedly tropical in their preferences (Chromodorididae, Hexabranchidae). The Corambidae, Goniodorididae, Onchidorididae, Polyceridae, Cadlinidae, and Archidorididae are families that seem to be centered chiefly on the cool northern temperate regions.

The Doridacea often possess rhinophoral cavities into which the tentacles may be retracted, but they only rarely have external sheaths surrounding their bases. The anal papilla is nearly always situated mid-posteriorly, under the mantle rim in *Corambe* and *Doridella* (= *Corambella*) and, incidentally, in all dorids at an early developmental stage, but situated dorsally in all the typical forms, such as *Doris*, *Archidoris*, *Polycera*, *Adalaria*, *Hexabranchus*, and many others. A variable number of retractile foliaceous gill plumes emerge from the mantle surface close to the anal papilla. In typical dorids the gill plumes are arranged in a circlet. They may be retracted in a coordinated way into a dorsal pocket on alarm in *Archidoris*, *Discodoris*, *Jorunna*, *Chromodoris*, *Asteronotus*, and many other dorid genera. This pocket may, in *Asteronotus cespitosus* from the South China Sea, be distinctly crenulate so the lips interlock and give greater resistance to an inquisitive fish. In less advanced dorids, such as *Laila*, *Polycera*, *Crimora*, *Adalaria*, and *Hexabranchus*, no such pocket is present and each gill is contracted separately down to the mantle surface on alarm. Sometimes large spiculose mantle papillae give extra protection to the gills and rhinophores, often containing elaborate defensive glands and luminescent or colored lures to draw the attention of a predator away from the fragile vital parts. Such papillae never contain lobes of the digestive gland in the Doridacea, and the liver usually forms a more or less compact single mass close to the stomach. The radula is usually rather broad and certain-

Above: The flamboyant *Okenia elegans* buries itself deep within its ascidian prey and seldom moves about if food is available. Such a life style is reminiscent of parasitism. *Below:* Dendrodorids like this *Dendrodoris subpellucida* are soft and slimy and lack a radula. Their feeding habits are unknown.

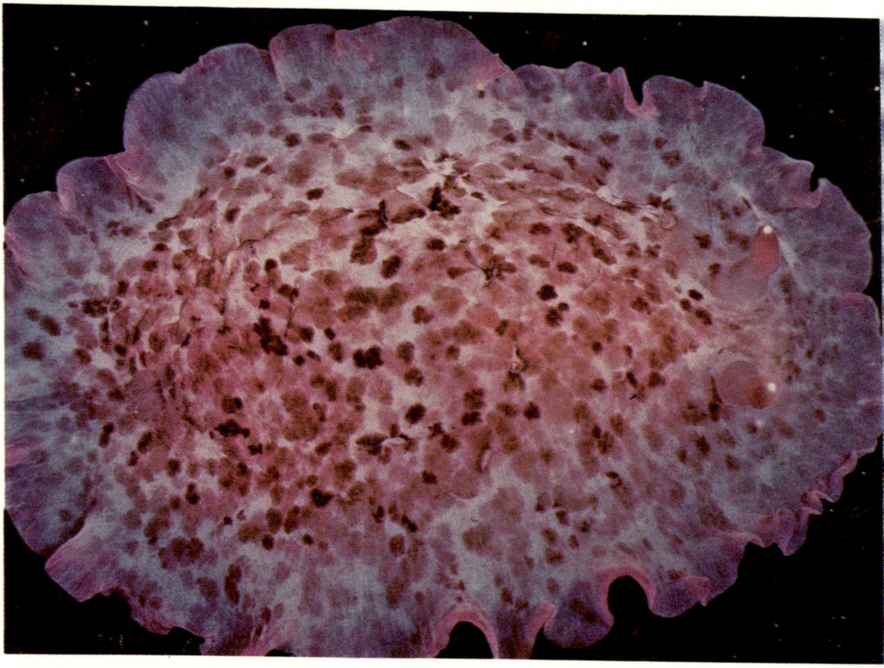

In the laboratory the doridacean *Triopha carpenteri* might appear, with its orange-red-tipped papillae, to show typical warning coloration. But when seen feeding on its prey, an encrusting orange sponge, its colors serve to conceal it.

Doridacean radulae. *Above: Goniodoris nodosa*, feeds on compound ascidians. *Below left: Okenia elegans*, feeds on simple ascidians. *Below right: Kalinga ornata*, feeds on sponges.

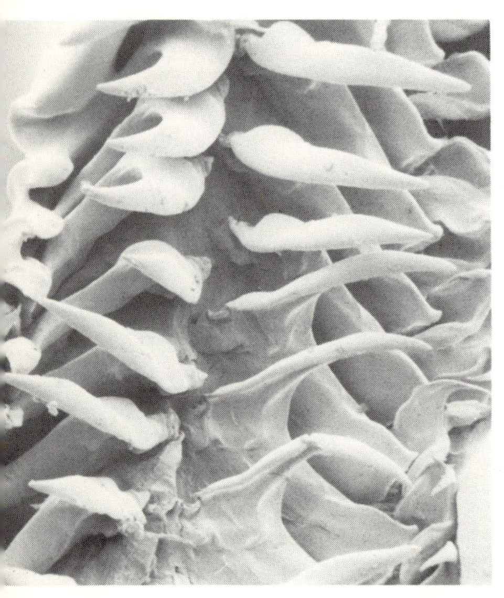

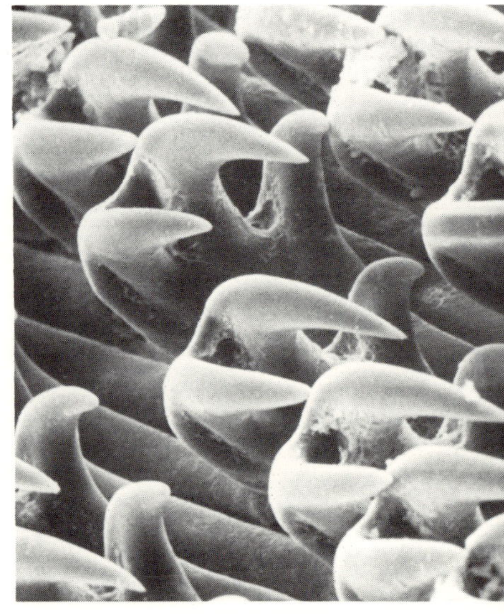

ly never uniseriate. One successful family of dorids, the Dendrodorididae, contains species which lack the radula and jaws and ingest sponges by an ingenious sucking modification of the muscular buccal mass. Sponges, bryozoans, acorn barnacles, compound sea squirts, and tube-dwelling polychaete worms form the diet of most dorid nudibranchs.

FAMILY	REPRESENTATIVE GENERA
Doridoxidae	*Doridoxa*
Bathydorididae	*Bathydoris*
Corambidae	*Corambe, Doridella (= Corambella)*
Goniodorididae	*Goniodoris, Okenia, Ancula, Trapania, Hopkinsia*
Onchidorididae	*Onchidoris, Adalaria, Acanthodoris*
Triophidae	*Triopha, Plocamopherus, Crimora, Kalinga*
Notodorididae	*Notodoris, Aegires*
Polyceridae	*Polycera, Thecacera, Laila, Limacia*
Gymnodorididae	*Gymnodoris, Nembrotha*
Vayssiereidae	*Vayssierea, Okadaia*
Hexabranchidae	*Hexabranchus*
Cadlinidae	*Cadlina*
Chromodorididae	*Chromodoris, Hypselodoris, Casella, Miamira, Ceratosoma*
Actinocyclidae	*Actinocyclus, Hallaxa*
Aldisidae	*Aldisa*
Rostangidae	*Rostanga*
Dorididae	*Doris, Austrodoris, Alloiodoris*
Archidorididae	*Archidoris, Trippa, Atagema*
Homoiodorididae	*Homoiodoris*

Above: The gaudy doridacean *Thecacera pennigera* is now known to be locally common and to feed on the bryozoan *Bugula plumosa*. The white-tipped clubs are filled with repellent glands. *Below:* The outlines of the internal organs may be seen through the flesh of these *Polycera faeroensis*. This doridacean feeds on bryozoans.

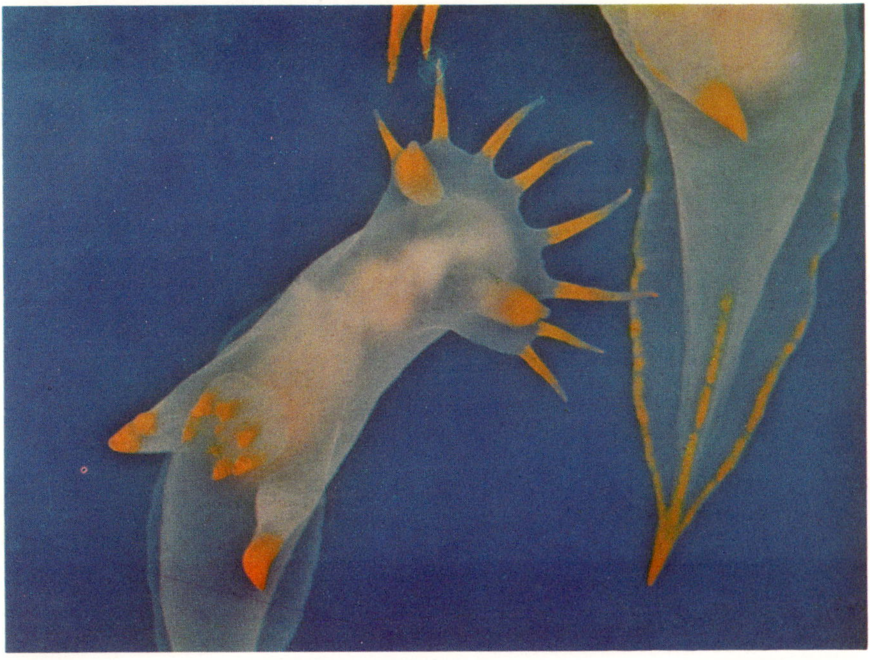

Polycera tricolor and other members of the genus have brilliantly marked defensive papillae situated so as to ward off attacks on the head and the delicate gills.

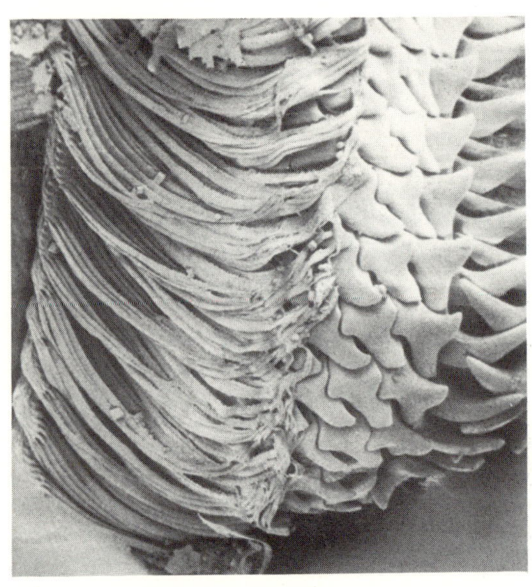

Doridacean radulae. *Above: Crimora papillata*, feeds on bryozoans. *Below left: Acanthodoris nanaimoensis*, feeds on bryozoans. *Below right: Poly-cera capensis*, feeds on bryozoans.

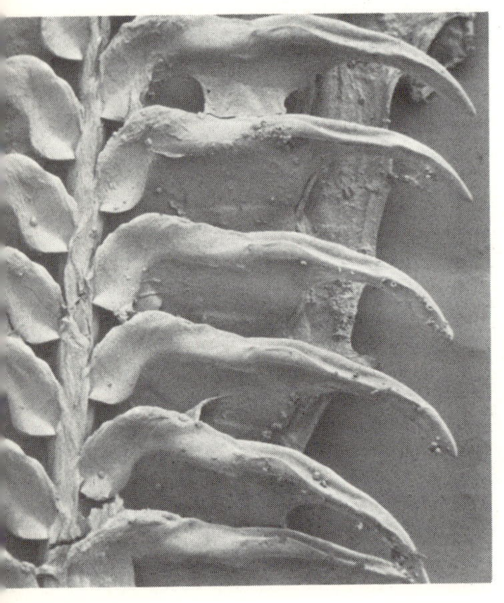

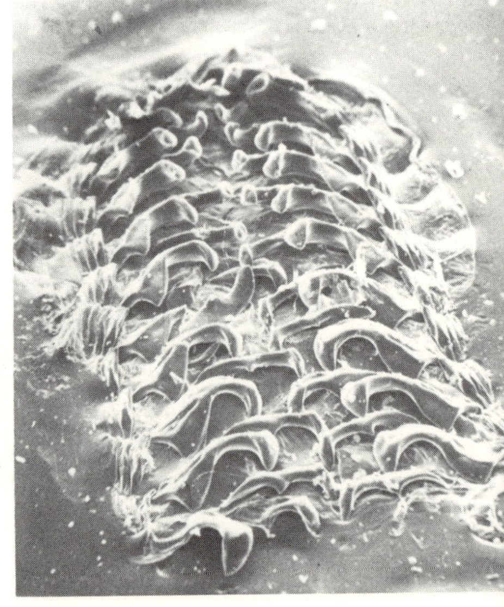

Doridacean radulae. *Above: Rostanga arbutus*, feeds on sponges. *Below left: Chromodoris amoena*, feeds on sponges. *Below right: Cadlina modesta*, feeds on sponges.

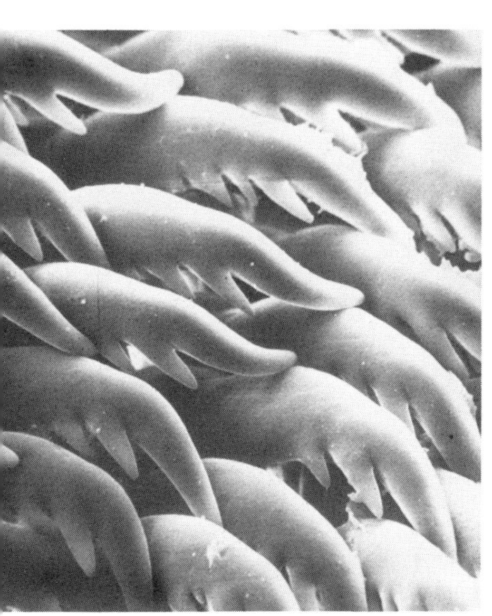

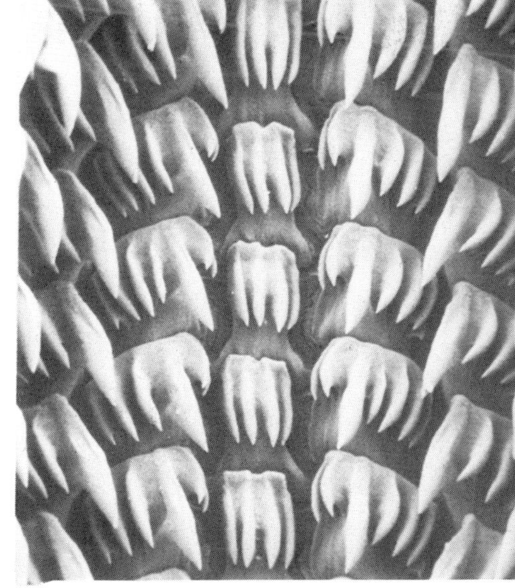

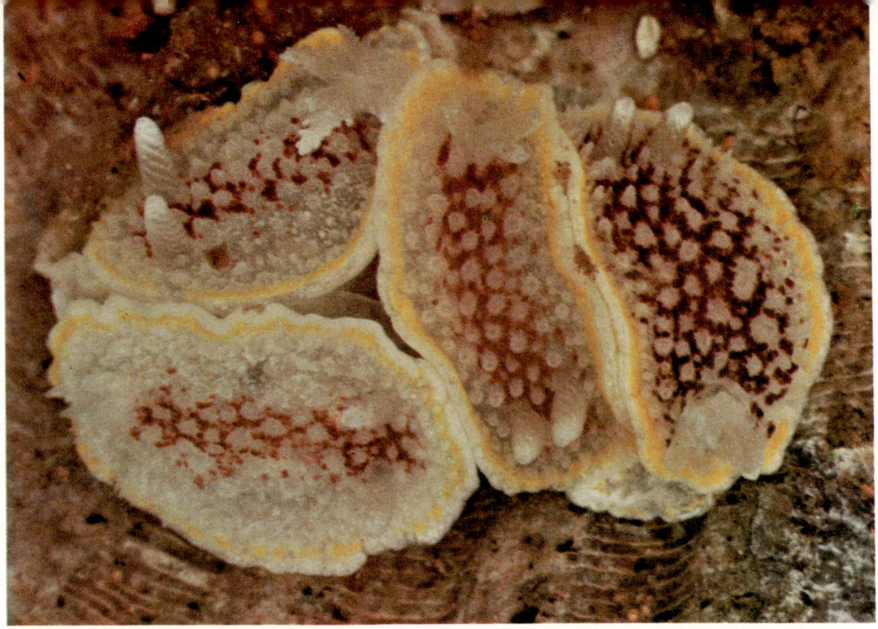

Above: During breeding season gregarious associations occur, probably in response to chemical attractants. Mating occurs in a head-to-tail fashion, each individual being hermaphrodite. Shown here is *Onchidoris luteo-cincta. Below:* The plainly colored *Adalaria proxima* on its prey bryozoan, *Electra pilosa.* This species has been extensively studied.

Eastern Australian doridaceans are probably the most beautiful of all the nudibranchs. This is *Chromodoris amoena,* an extraordinarily variable species; the pattern of spots and blotches is never the same in two individuals.

Baptodorididae	*Baptodoris*
Discodorididae	*Discodoris*
Kentrodorididae	*Kentrodoris, Jorunna*
Asteronotidae	*Asteronotus, Sclerodoris, Halgerda, Aphelodoris*
Platydorididae	*Platydoris, Hoplodoris*
Phyllidiidae	*Phyllidia*
Dendrodorididae	*Dendrodoris*

Suborder III. ARMINACEA

This suborder is more difficult to define and certainly more difficult to recognize from external features alone. The nine families currently contained here will not fit into any other suborder, which (to a certain extent) explains our conviction that a separate suborder is needed for them. Species of *Armina* found in the North Atlantic have, externally, some of the features of a primitive doridacean such as *Phyllidia* or *Corambe*, with external pallial leaflets forming a respiratory series beneath the mantle edge, but internally *Armina* proves to be very unlike a dorid, for it has a much-divided liver, a laterally placed anal papilla, and strong jaws.

Similarly, *Hero, Antiopella*, and the Pacific Ocean *Dirona* have a strong resemblance to the aeolidacean nudibranchs which will be dealt with shortly. But their dorsal ceratal processes lack cnidosacs with contained nematocysts, a fact which instantly separates them from the aeolids.

The arminacean nudibranchs do, on the other hand, have certain features in common. Their rhinophoral tentacles do not possess external protective sheaths. The anal papilla is usually situated rather far forward, either dorsally or laterally, on the right side. The radula may be narrow, but it is never uniseriate (with a single tooth in each row). Oral tentacles are usually lacking.

The diets of the arminaceans are varied. *Hero formosa* certainly feeds upon naked or gymnoblastic hydroids, and *Armina*, too, is said to attack coelenterates (the alcyonarian sea pansies in the case of the best-known species, *Armina californica*). But *Antiopella* and *Proctonotus* feed upon encrusting and erect bryozoans, while *Dirona albolineata* is known to devour a wide variety of shelled mollusks and other invertebrates.

FAMILY	REPRESENTATIVE GENERA
Heterodorididae	*Heterodoris*
Doridomorphidae	*Doridomorpha*
Arminidae	*Armina, Linguella, Dermatobranchus*
Madrellidae	*Madrella*
Dironidae	*Dirona*
Antiopellidae	*Antiopella*
Gonieolididae	*Gonieolis*
Charcotiidae	*Charcotia*
Heroidae	*Hero*

Suborder IV. AEOLIDACEA

This is a very compact group containing 20 families of delicate, beautifully formed and patterned species. They all bear clusters, groups, or rows of elongated finger-like smooth dorsal ceratal processes. These cerata contain the much-divided lobes of the digestive gland, together with, at their tips, the defensive cnidosacs. The cerata are usually vividly marked in a characteristic way in each species. These cerata are extremely short in the sand-burrowing aeolid *Pseudovermis* but in the majority of the species are the most conspicuous feature of the body. The rhinophoral tentacles are never retracted into pallial sheaths, while the oral tentacles are often very long and graceful and the propodial extremities are sometimes extended so as to form a third pair of anteriorly placed sensory processes.

Jaws are usually well developed, but the radula is reduced to a very narrow ribbon and sometimes bears only a

Above: Doridella steinbergae is hard to detect in this photo of three specimens on their food, the bryozoan *Membranipora*. *Below:* The brilliantly colored ceratal tips of this aeolidacean, *Trinchesia caerulea,* draw the attention of predators away from the defenseless head.

Nowhere in the world is there a better display of aeolidacean nudibranchs than in the Tyrrhenian Sea, an arm of the Mediterranean. *Coryphella lineata* (white line down back), *Coryphella pedata* (magenta), and *Hervia costai* (red spots on head) are among the most beautiful of all marine animals.

AEOLIDACEAN RADULAE.

Facelina auriculata coronata, feeds on hydroids.

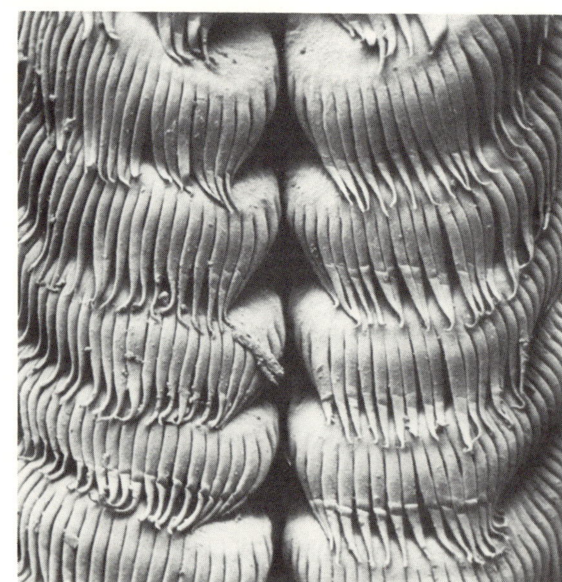

Aeolidiella glauca, feeds on sea anemones.

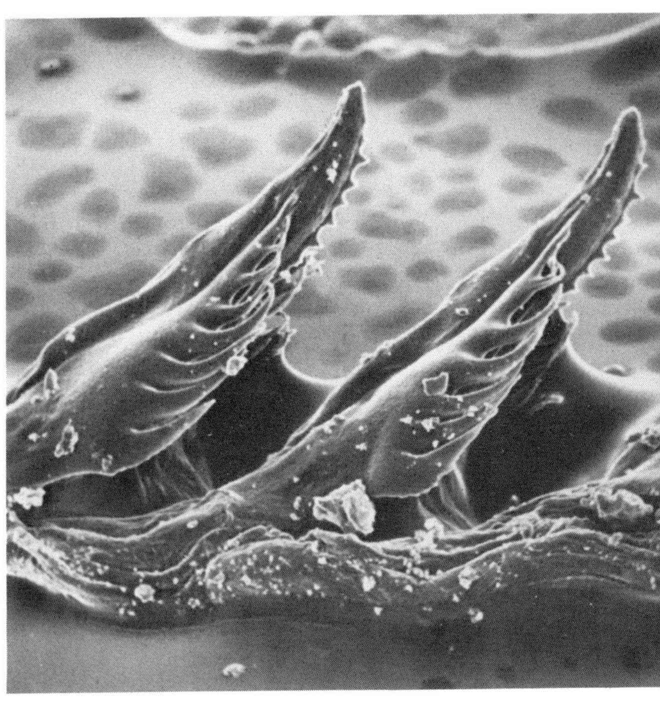

Aeolidacean radula: *Hermissenda crassicornis*, feeds on hydroids.

single file of teeth (the uniseriate condition). Aeolidaceans (such as *Aeolidia*, *Eubranchus*, and *Catriona*) are usually active hunters feeding upon coelenterates. *Glaucus* and *Glaucilla* are planktonic aeolids preying upon siphonophores and chondrophores. The Pacific American species *Phidiana pugnax*, however, attacks other opisthobranchs, even some aeolids, as its normal prey, while *Calma* and some species of *Favorinus* devour only the eggs of their molluscan and fish prey. *Favorinus ghanensis*, however, feeds on the bryozoan *Zoobotryon verticillatum* in West Africa. *Fiona pinnata* is a very unusual aeolid of wide geographical distribution which occurs occasionally in American waters and attacks stalked barnacles such as *Lepas*.

FAMILY	REPRESENTATIVE GENERA
Notaeolidiidae	*Notaeolidia*
Coryphellidae	*Coryphella*

Above: Favorinus blianus feeds on nudibranch eggs instead of coelenterates and thus cannot utilize nematocysts in defense. Its color varies with the food taken; these specimens have fed on orange and white eggs. *Below:* Aeolidaceans are aggressive and fast-moving nudibranchs, and *Facelina annulicornis* is one of the more aggressive species. Aeolidaceans should not be crowded or they become cannibalistic. (The spawn band is of a *Doto*.)

Pteraeolidia semperi is an Indo-Pacific aeolidacean nudibranch in which the dorsal papillae (cerata) are held in a series of fan-like bunches. This is the largest aeolidacean known from Australian waters and may grow up to 15cm in length.

Nossidae	*Nossis*
Protaeolidiellidae	*Protaeolidiella*
Caloriidae	*Caloria*
Phidianidae	*Phidiana, Moridilla*
Facelinidae	*Facelina, Hermissenda*
Favorinidae	*Favorinus*
Myrrhinidae	*Myrrhine, Phyllodesmium*
Glaucidae	*Glaucus, Glaucilla*
Pteraeolidiidae	*Pteraeolidia*
Herviellidae	*Herviella*
Aeolidiidae	*Aeolidia, Aeolidiella, Cerberilla*
Spurillidae	*Spurilla, Berghia*
Eubranchidae	*Eubranchus, Egalvina*
Cumanotidae	*Cumanotus*
Flabellinidae	*Flabellina*
Pseudovermidae	*Pseudovermis*
Cuthonidae	*Precuthona, Cuthona, Phestilla, Catriona, Tenellia, Tergipes, Embletonia*
Fionidae	*Fiona*
Calmidae	*Calma*

Locomotion

Many opisthobranchs move about little while they are undisturbed and food is abundant. They can move, however, with surprising speed if they are starved or alarmed by an enemy. Most nudibranchs spend the greater part of their lives creeping or burrowing (sometimes deep within their food), but swimming movements can be spectacular, especially in the energetic *Hexabranchus* of the Coral Sea or the sculling aeolidacean of the American Pacific coast, *Cumanotus beaumonti*.

Creeping in nudibranchs is brought about by finely controlled movements of the interlacing longitudinal and dorso-ventral smooth muscle fibers which pack the foot. Local areas of the foot can be raised from the substrate and moved to a more forward position. These minute movements are usually coordinated so that transverse waves of activity can be seen (if an individual creeping on a glass plate is observed) moving over the pedal sole. In most recorded cases the transverse waves are retrograde, that is to say, they travel backward over the sole, contrary to the direction of locomotion.

Swimming is only possible in mollusks in which the shell has become flimsy, buoyant, internal, or altogether lost. It seems probable that the evolutionary pressure which led to the adoption of swimming in so many groups of opisthobranch mollusks was the need to escape from slow-moving bottom-living predatory carnivores. Many starfish and some crustaceans, for example, will eat virtually any animal materials which remain motionless for very long. A number of nudibranchs possess swimming escape reactions designed to minimize these dangers.

A large nudibranch of the American Pacific coast, *Tritonia diomedea*, swims away if a large specimen of the carnivorous many-armed starfish *Pycnopodia helianthoides* is

Above: *Precuthona peachi* is an aeolidacean which feeds only on the hydroid *Hydractinia* growing on shells occupied by hermit crabs. The animal and its spawn masses are both well camouflaged. *Below:* Aeolidaceans are more agile than most other nudibranchs. This *Coryphella pedata* is righting itself after being disturbed.

Calmella cavolinii is a small and delicate aeolidacean which rarely reaches a body length of 10mm. The body is so slender that it is a wonder that the complex viscera can be fitted within it.

placed nearer to it than a couple of feet in still sea water. The *Tritonia* springs off the bottom by convulsive dorso-ventral muscular contractions of the whole body. Such movements are not very efficient and are continued for only a minute or two. The strong water currents in the habitat preferred by *Tritonia* will normally be enough to move it away from danger. During the recovery stroke the animal may drop through the water almost as much as it gained on the last effective (i.e., sole-flexing) stroke. The only conspicuous modification to aid in swimming in *Tritonia diomedea* is the spatulate metapodium (rear of the foot), which is flattened dorso-ventrally.

Other nudibranchs may employ lateral undulations of the whole body in order to achieve the same end. *Dendronotus iris*, a common nudibranch of the American Pacific coast, will react in this way to rough handling. In converting the side-to-side movements of the body into forward translocation, the expanded, trumpet-like rhinophore sheaths play an important part. These organs are enlarged in the dendronotacean nudibranchs and in *D. iris* are used as oars during the swimming excursions. The swimming mechanism of the highly specialized suspension-feeding sea-slug *Melibe leonina* is rather similar, and the role of the dilated rhinophore sheaths is identical. The aeolidacean *Coryphella iodinea* and the dorid *Nembrotha eliora* also swim by lateral body movements.

The Indo-Pacific dorid nudibranch *Plocamopherus ceylonicus* has been seen to swim by lateral flexions of the body, approximately one per second. Each violent contraction of the longitudinal musculature of the body brings the head and tail-tip together. During swimming the oral veil is spread out widely, the rhinophores are held back along the dorsum, and the gills are extended. Two postero-dorsal crests or keels, one formed by the mantle and the other by the metapodium, are erected and serve to increase the effectiveness of the clumsy swimming movements. Using this mechanism, *Plocamopherus* can rise a few inches from the bottom. Lift is achieved at each stroke by bringing the metapodial keel to a position *below* the oral veil. The swimming move-

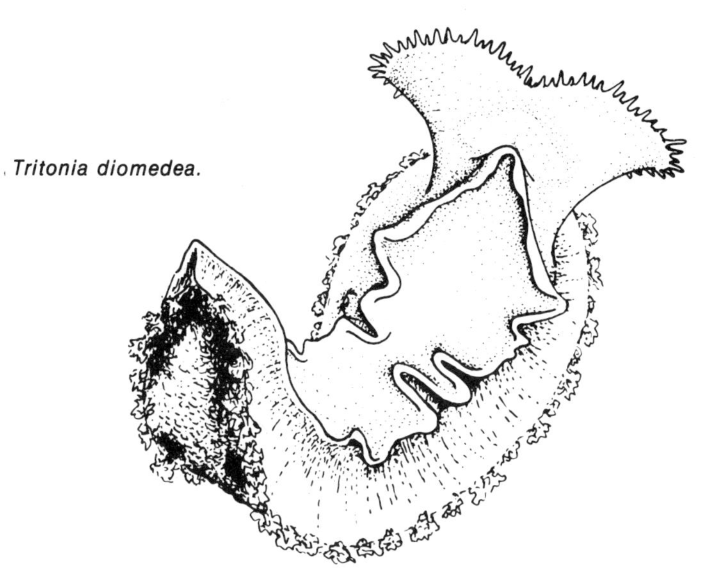

Tritonia diomedea.

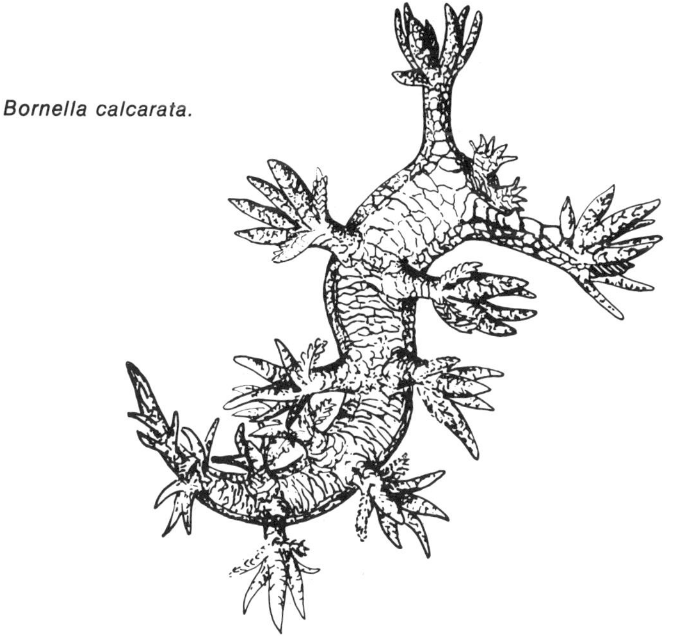

Bornella calcarata.

Above: Cumanotus beaumonti is an aeolidacean which swims by vigorous up and down movements of the cerata. *Below:* The pelagic *Glaucus atlanticus* is found in all tropical areas, where it feeds on the pelagic coelenterates *Physalia*, *Velella*, and *Porpita*. It is able to utilize *Physalia* nematocysts, making it potentially dangerous to humans.

Despite their alertness and general aggressiveness, aeolidaceans do get hurt. This *Facelina auriculata coronata* has suffered damage to one of the oral tentacles. Although it has undergone regeneration, in this unusual instance an abnormal forked tip has been produced.

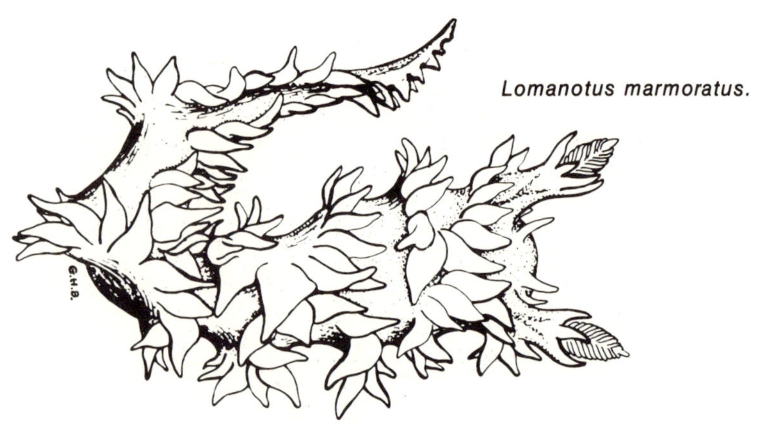

Lomanotus marmoratus.

ments of the dendronotacean nudibranchs *Scyllaea pelagica* and *Bornella digitata* are rather similar to those of *Plocamopherus*, but these species often swim upside down.

The aeolidacean nudibranch *Cumanotus beaumonti* escapes from enemies in a rather different way. In this widely distributed sea-slug the cerata undergo up and down movements like those of a pigeon's wings, pivoting at their bases. The arrangement of the striated muscles which bring this about has not been fully investigated. When the animal is abruptly disturbed, repeated downward sweeps of the cerata propel the body away from the bottom for minutes at a time.

From humble beginnings like these, the varied swimming mechanisms found in the Nudibranchia arose. When a swimming reaction had evolved for the purpose of escaping from predators, in many cases it became further perfected in connection with a secondary function such as aggregation for breeding or feeding in the surface layers of the sea. There is much diversity in the structures utilized as swimming organs.

Let us consider some nudibranchs in which swimming is not primarily an escape response (although here, too, an abrupt disturbance will often initiate swimming movements). Few of these utilize whole-body flexions; only the leaf-like *Phylliroe* and a few allies among the nudibranchs swim perpetually by lateral waves of contraction passing back from the head along the body. This nudibranch is so

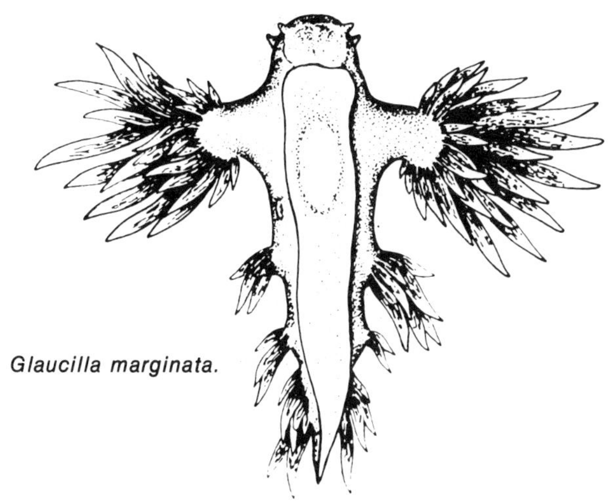

Glaucilla marginata.

perfectly adapted for its pelagic life that the creeping sole has been lost in the adult.

The aeolidaceans *Glaucus atlanticus* and *Glaucilla marginata* manage to stay in the surface layers of the sea (where they feed upon the planktonic coelenterates *Velella*, *Porpita*, and *Physalia*) by a novel behavioral modification. They gulp in air from above the water surface and hold a bubble inside the stomach. If the nudibranchs are squeezed so as to expel the bubble, they sink to the bottom. Glaucid nudibranchs are passively planktonic. Movements of parts of the body have been reported, but these do not appear to result in translocation.

Tethys fimbria.

Aeolidia papillosa is one of the largest aeolidacean nudibranchs, and it is found on both sides of the Atlantic. The numerous cerata on the back contain nematocysts derived from prey sea anemones. Shown below is a single erupted nematocyst and several unerupted cells. The barbed shaft or thread is easily visible.

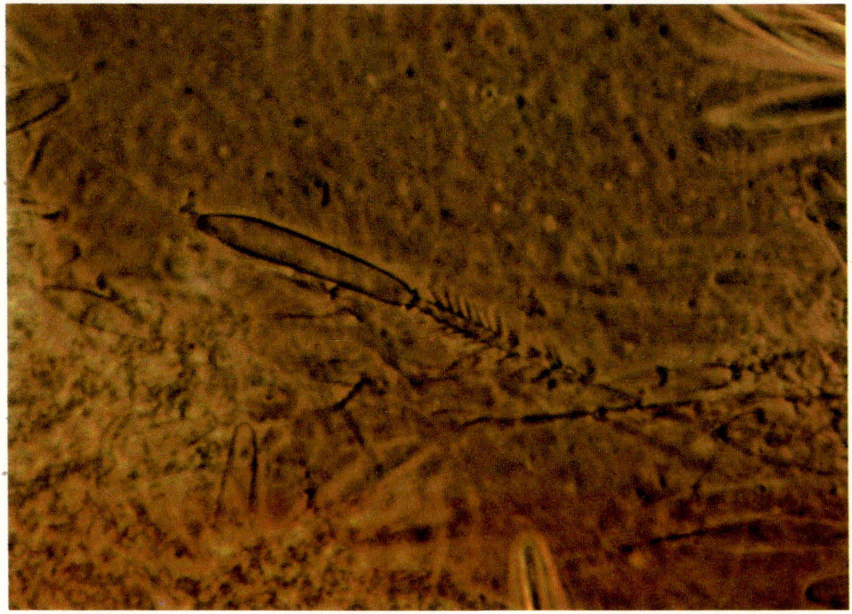

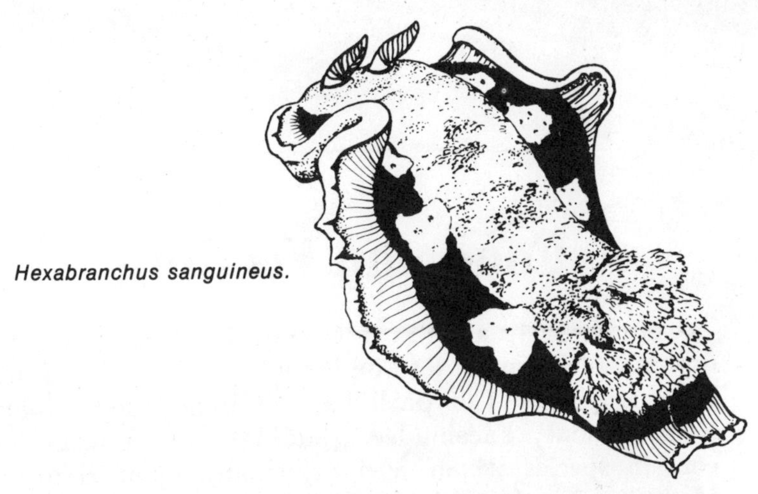

Hexabranchus sanguineus.

The splendidly marked coral reef doridacean *Hexabranchus sanguineus* swims in a fashion which is unique among gastropod mollusks and finds parallels only among certain cephalopods. When at rest the margins of the mantle skirt are dorsally enrolled and slow locomotion occurs by creeping on the broad pedal sole. In swimming, however, the mantle skirt is spread out and strong locomotor waves are propagated rearward from the anterior pallial margins. The waves are synchronous on the two sides of the body, and each wave takes about four seconds to travel its course. Two waves are usually in progress along the sides of the animal at any moment. At the same time the whole body undergoes great dorso-ventral flexions, a full cycle of activity occupying about four seconds. Swimming may continue for many minutes, and it is probable that this is more than a simple escape reaction. Certainly a brilliant display of color accompanies the unrolling of the mantle edge, and it may be that swimming here is the behavioral component of an aposematic or warning display. While swimming, the lateral edges of the foot are brought together in order to conceal the sole and the rhinophores are held back against the dorsal mantle. These behavioral modifications, combined with the sudden revelation of the brilliantly marked rim, may well serve to alarm a predator.

Food and Feeding

Nudibranchs are all carnivorous. Each family contains species which usually feed on broadly similar types of prey. For example, the Coryphellidae, Dendronotidae, Dotoidae, Eubranchidae, Facelinidae, Heroidae, and Lomanotidae contain species which feed principally upon Hydrozoa. Many species of the Aeolidiidae feed upon Actiniaria (although the American *Phidiana pugnax* feeds chiefly upon other aeolid nudibranchs), and the Glaucidae all attack chondrophores and siphonophores. Members of the Tritoniidae all feed on alcyonarian corals. The Antiopellidae and Onchidorididae contain species which attack bryozoans; *Onchidoris bilamellata*, however, feeds upon sessile barnacles.

A feeding type which has proved rather successful in members of several opisthobranch families is the egg-consumer. This habit crops up in the aeolids, such as *Favorinus branchialis* and *Calma glaucoides*, as well as in the 'false nudibranchs' or sacoglossans *Stiliger vesiculosus* and *Olea hansineensis*. Mollusks like these are often very pale in color so as to pass unnoticed among cream-colored or white eggs. Whereas *Calma* feeds on the eggs of cephalopods and teleost fishes, *Favorinus*, *Stiliger*, and *Olea* take the eggs of other opisthobranchs (up to 20 eggs per minute). On examining such predators, it is not uncommon to detect not only ova and young cleavage stages in the stomachs, but also late veligers of the molluscan prey species. In *Calma* the anal opening is closed off in post-larval stages, presumably because there are few solid waste materials remaining after the digestion of such yolky food. *Olea* and *Favorinus* are less firmly committed to a diet of embryos, and in these genera the anus remains functional throughout free life. The West African *F.*

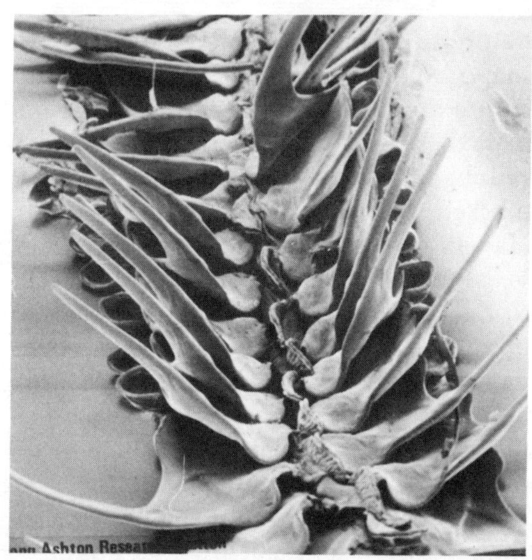

Doridacean radula: *Onchidoris bilamellata,* feeds on acorn barnacles.

ghanensis, moreover, does not feed on eggs but instead crops the polyps of the bryozoan *Zoobotryon verticillatum.*

Another highly successful feeding specialization has evolved independently in the Fionidae (*Fiona*) and the Glaucidae (*Glaucus* and *Glaucilla*). It involves attacking the planktonic cnidarians *Porpita, Physalia,* and *Velella. Fiona* is said to also attack stalked barnacles. A recent report described *F. pinnata* feeding on *Lepas anatifera* in Californian waters. These prey organisms are extremely common in the oceans, and glaucid and fionid nudibranchs are found in all warm seas. Both *G. atlanticus* and *F. pinnata* are circumtropical in distribution.

Many dorid nudibranchs are well adapted for feeding upon encrusting marine sponges and possess dorso-ventrally flattened bodies, a more or less oval outline, and a broad, flattened foot. Such a body form is exhibited also by other nudibranchs which browse on flattish encrustations (for example, *Alcyonium,* bryozoan and ascidian colonies, and barnacle aggregations). Their radula is usually quite broad and, with a few conspicuous exceptions among the pleurobranchomorphs and the tritoniids, lacks strong jaw plates. In nature they are frequently well camouflaged through their shape, coloration, and behavior. Mysteriously, these remarks do not apply well to the chromodorid nudibranchs of warm seas; these feed upon encrusting sponges and possess a broad radula and feeble jaws but are often extraordinarily vividly marked. Nor do they apply to *Okadaia elegans*, which in Hawaii drills through the calcareous tubes of spirorbid and serpulid polychaetes so as to consume the worms within, or to the dendrodorid nudibranchs of warm seas which feed upon sponges by the novel method of sucking up the sponge tissues through a strongly developed proboscis (the radula is absent).

Nudibranchs which feed upon arborescent, more erect organisms (for example, hydroids, actiniarians and some bryozoans) are usually of smaller size and of more elongate shape, with a long narrow foot adapted for clinging to this type of prey. Nudibranchs which fall into this category are active, voracious forms in which cannibalism may often be seen in laboratory aquaria. Jaws are often well developed, and the radula shows a tendency toward reduction in width. In extreme cases (*Aeolidia, Hermissenda, Pteraeolidia*) the radula may exhibit only one (central) tooth in each row. The defenses of animal prey frequently include internal or external spicular concretions or an exoskeletal refuge into which the soft parts may be rapidly retracted, the aperture often being furnished with an opercular system. The prey may further be rendered unpalatable by its possession of nematocysts (Cnidaria) or a strongly acidic internal fluid (Ascidiacea).

In overcoming these defenses, the secretion by the nudibranch of great quantities of mucus from oral, propodial, and other glands is of importance. Ingested food is liberally covered and mixed with mucus as a protection against abrasion. Despite this, it is recorded that sponge spicules often pierce the stomach wall of *Archidoris pseudoargus*. To manage the food, the nudibranch is provided with a muscular buccal mass containing the radula and often a pair of jaws. This apparatus may be situated at the inner end of an inverted oral canal so that when extended the radula and jaws come to form the tip of a mobile proboscis (as in *Kalinga*).

The mode of functioning of the components of the buccal apparatus varies greatly in different species. In many species of dorid nudibranchs (for instance *Archidoris pseudoargus* and *Jorunna tomentosa*) jaws are absent and the prey is scooped up, fragmented, and passed into the stomach solely by the action of the broad radula and its musculature. *Tritonia plebeia*, on the contrary, possesses a broad radula which is employed only to transport intact pieces of the prey into the stomach after they have been sliced away from an alcyonarian by the action of the powerful jaws.

By the action of the buccal mass the opercular plates of barnacles, in the case of *Onchidoris bilamellata*, or of bryozoans, in the case of *Adalaria proxima* or *Doridella* (= *Corambella*) *steinbergae*, may be broken open and the soft parts sucked out by rhythmic dilations of a muscular buccal diverticulum; pieces of *Alcyonium* more than a centimeter across may be cut cleanly away and forced into the stomach (in *Tritonia hombergi*); calyptoblastic hydroids may be cropped and tubularians decapitated by facelinids, eubranchids, and other nudibranchs.

Whereas some nudibranchs (for instance *Archidoris pseudoargus* and *Tritonia hombergi*) kill and ingest their prey with little waste, others (for instance *Facelina auriculata* and *Aeolidia papillosa*) often kill more of the prey than they eat, and wastage may be considerable. It is of interest that *Archidoris* and *Tritonia* have a highly restricted diet, while *Facelina* and *Aeolidia* are known to feed on a variety of cnidarians. In the former cases dietary specialization has clearly been paralleled by a higher manipulative efficiency.

In many nudibranchs, for instance *Archidoris pseudoargus* and *Tritonia hombergi* (which are, incidentally, two of the largest North Atlantic nudibranchs), the prey may consist of a single species in each case. Observations throughout the year on *A. pseudoargus* showed its complete dependence through all the benthic stages of the life cycle on *Halichondria panicea*. The dorids were nearly always found on an encrustation of this sponge and were often partially embedded in it. This close dependence upon a single prey species is also found in the Mediterranean dorid *Peltodoris atro-*

maculata, which feeds only on the sponge *Petrosia dura*. *Tritonia hombergi* is almost invariably found by divers in association with the soft coral *Alcyonium digitatum*. The related *T. plebeia*, however, is known to feed both on *Alcyonium* and *Eunicella verrucosa*.

In most of the nudibranchs diet specificity is not so rigid and many species are known to take a variety of prey. *Doto fragilis* and *Eubranchus exiguus*, for example, feed on several species of calyptoblastic hydroids; *Acanthodoris pilosa* has been seen to attack a number of encrusting intertidal bryozoans. Several species of nudibranchs have been recorded feeding on a large number of species of food organisms. *Doto coronata* has been found to attack more than twenty species of hydroids (mainly thecate forms). The

Doridacean radula: *Triopha carpenteri*, feeds on sponges.

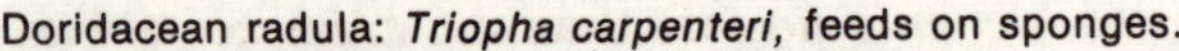

Pacific American arminacean *Dirona albolineata* may ingest small living snails (*Lacuna carinata, Margarites pupillus,* and *M. helicinus*) in a locality 15m or less in depth near San Juan Island, but in greater depths feeds less selectively, scraping up ectoprocts, hydroids, small crustaceans, sponges, barnacles, tunicates, diatoms, coralline algae, and detritus. In the case of very few nudibranch species have preferences been elucidated. One such case is the small doridacean *Adalaria proxima*, common intertidally around the British coast. If given a choice in the laboratory between the bryozoans *Alcyonidium polyoum, Electra pilosa, Flustrella hispida,* and *Membranipora membranacea*, these dorids almost invariably attack a colony of *E. pilosa*. On British shores they are usually found on this bryozoan, although all these species are abundant on the shore. In aquaria adult *Adalaria* will feed on the other species only if denied access to *Electra*.

Information about selectivity is available also for *Aeolidia papillosa*, a nudibranch which preys on actiniarians. It appears to prefer *Actinia equina* to other anemones, although recent reports claim that *Anemonia sulcata* and *Anthopleura elegantissima* are also favored and young ones could be reared successfully on a diet of *Actinia* or of *Metridium senile*, but not on diets of *Diadumene cincta, Sagartia troglodytes,* or *Sagartiogeton undata*. In laboratory tests designed to analyze their chemosensitivity, adult *Aeolidia* were attracted at a distance to *Actinia* and *Metridium* but not to a number of other cnidarians. The case of *Aeolidia* illustrates that all known components of a nudibranch's diet may not be of equal palatability and nutritive content.

Taking another example, the Pacific American *Tritonia festiva*, we find evidence of differing dietary preferences in different parts of the nudibranch's geographical range. In Puget Sound *T. festiva* attacks the sea pen *Ptilosarcus guerneyi*, but around La Jolla, California, it feeds only on the pink gorgonian *Lophogorgia chilensis*. Somewhat surprisingly, it has been shown recently that specimens of *Aeolidia papillosa* from California performed similarly to British individuals in laboratory tests designed to show food preferences.

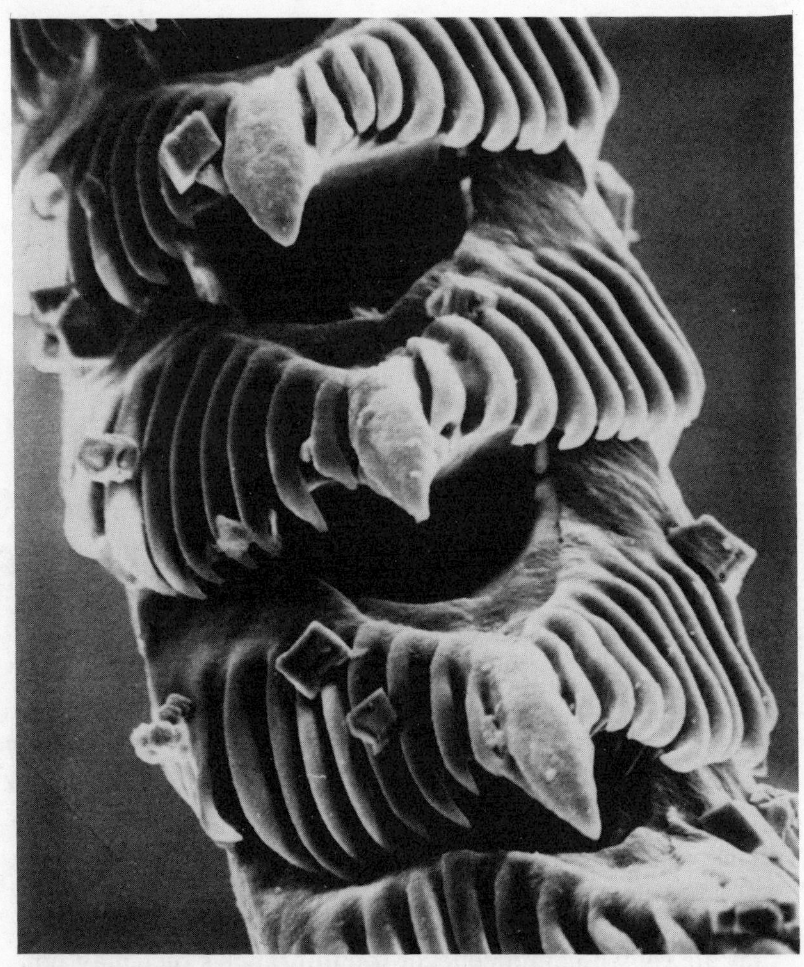

Aeolidacean radula: *Pteraeolidia semperi*, food unknown.

In a few cases it has been suggested that food preferences may alter with age. Young *Dendronotus frondosus* are abundant on *Sertularia cupressina* in the North Atlantic during the winter months, while older specimens (up to 5 cm in length) may be found on the shores and sea bottom feeding on *Tubularia indivisa* in spring and early summer. Similarly, while adult *Goniodoris nodosa* attack encrusting colonial ascidians, the young stages feed on the bryozoan *Alcyonidium polyoum* and perhaps on other bryozoan species.

Defense

The evolutionary loss of the most characteristic molluscan organ of mechanical and passive defense, the shell, brought with it several problems. Defense has become not mechanical, but chemical and biological; not passive, but active.

Aquarium experiments in which various live gastropods were fed to hungry fishes of several kinds showed that nudibranchs were almost invariably refused as food by fish. The tests were always carried out when the fish were expected to be hungry, and this was confirmed by giving them some of their normal food before the experiments. The nudibranch to be tested was then dropped into the tank from above and allowed to sink to the bottom. The usual reaction of the fish was to inspect the test organism closely; occasionally a fish would take a nudibranch into its mouth, but nearly always rejected it violently and immediately. In a tank containing several healthy pollack (*Pollachius pollachius*), a single test mollusk may be inspected, taken into the mouth, and rejected by every fish present before it reaches the tank bottom. The obvious health of the fish, their vigorous appetites, and, above all, their eagerness to take and swallow other food or dead or damaged nudibranchs showed that healthy nudibranchs are distasteful to fish.

We are beginning to understand the detailed functioning of some of the defensive mechanisms employed by nudibranch mollusks. First of all, it is important to recognize that any individual species may include in its repertoire several different defensive adaptations. For instance, *Discodoris planata* spends most of its life hiding under sponge-covered shallow water boulders in the North Atlantic. If nosed out by an inquisitive fish, it can in a crisis expel quantities of sulfuric acid from glands that open all over the skin. It has been known for many years that teleost fish have a common but inexplicable abhorrence of anything tasting acidic. In a paper published in 1890, the zoologist Bateson wrote that: "Conger are equally willing to eat a piece of squid or pilchard if it is covered or smeared with spirit, cheese of various sorts, anchovy extract, or *Balanoglossus* as if it had been unpolluted. On the other hand, they will refuse cooked or tainted food and food which has been soaked for a few moments in dilute acids. The remarks apply generally to the other fishes."

Another dramatic defensive adaptation found in many nudibranchs is the utilization of the nematocysts of the prey by many of those species that feed upon Cnidaria. The best known examples are found in the Aeolidacea. The manipulation and translocation of nematocysts in the alimentary canal of the aeolid nudibranchs are still mysterious processes. Certainly many cnidarian nematocysts are fired during the process of biting and swallowing, and examination of stomach contents of nudibranchs immediately after feeding showed as many as 50% of the nematocysts to have discharged prior to or during digestion. From the undischarged remainder a proportion are translocated through the digestive gland and into the cnidosacs within the dorsal papillae or cerata. There they become ingested by cnidophore cells which store and nourish them until they are required. This occurs when the nudibranch is attacked, as by a predatory fish. If any of the cerata are nipped, the nematocysts burst from their cnidophore cells and are expelled through the ceratal terminal pore. They discharge immediately when meeting sea water, and it may be imagined that the hungry fish often abandons the attack as a result. After discharging its nematocysts a ceratal cnidosac will remain empty until its normal prey is again encountered.

An interesting process of nematocyst selection has been described recently in the free-floating (planktonic) aeolidaceans of tropical seas, *Glaucus atlanticus* and *Glaucilla marginata*. These species both feed upon *Physalia*, *Velella*, or *Porpita* (all nematocyst-rich cnidarians), but retain for their own defense only the exceptionally potent nematocysts of

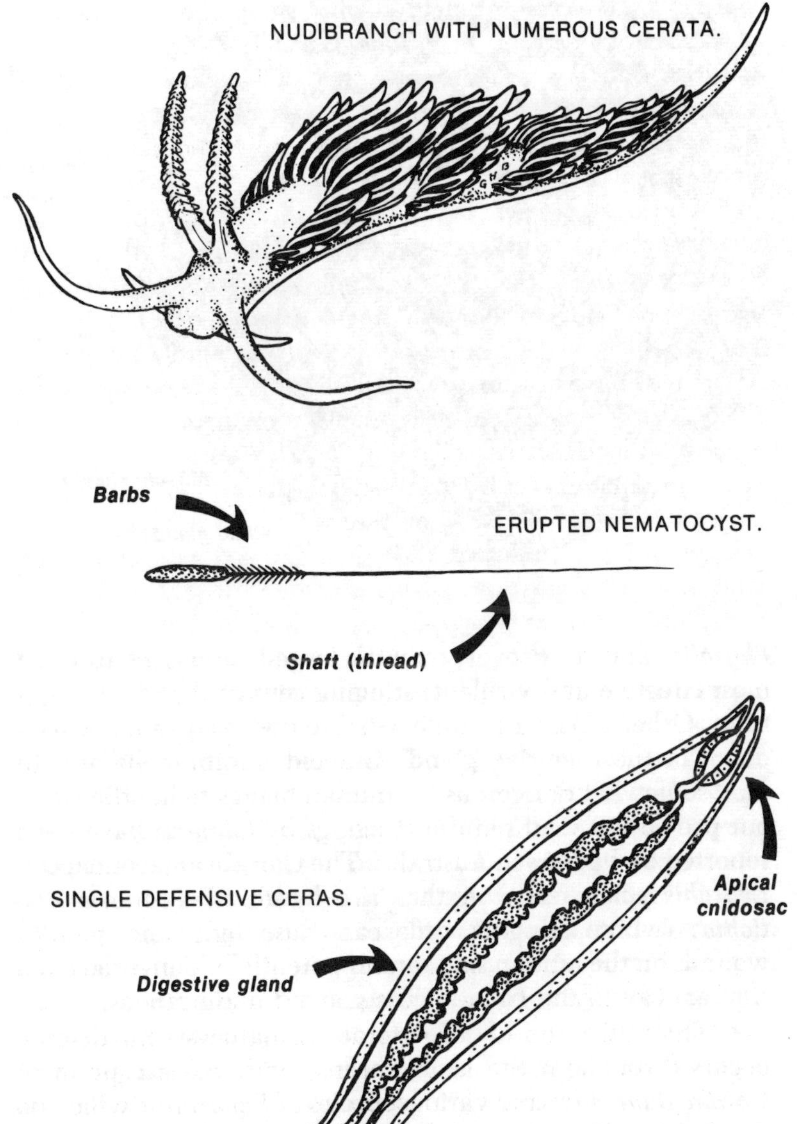

AEOLIDACEAN DEFENSE.
NUDIBRANCH WITH NUMEROUS CERATA.
Barbs
ERUPTED NEMATOCYST.
Shaft (thread)
SINGLE DEFENSIVE CERAS.
Apical cnidosac
Digestive gland

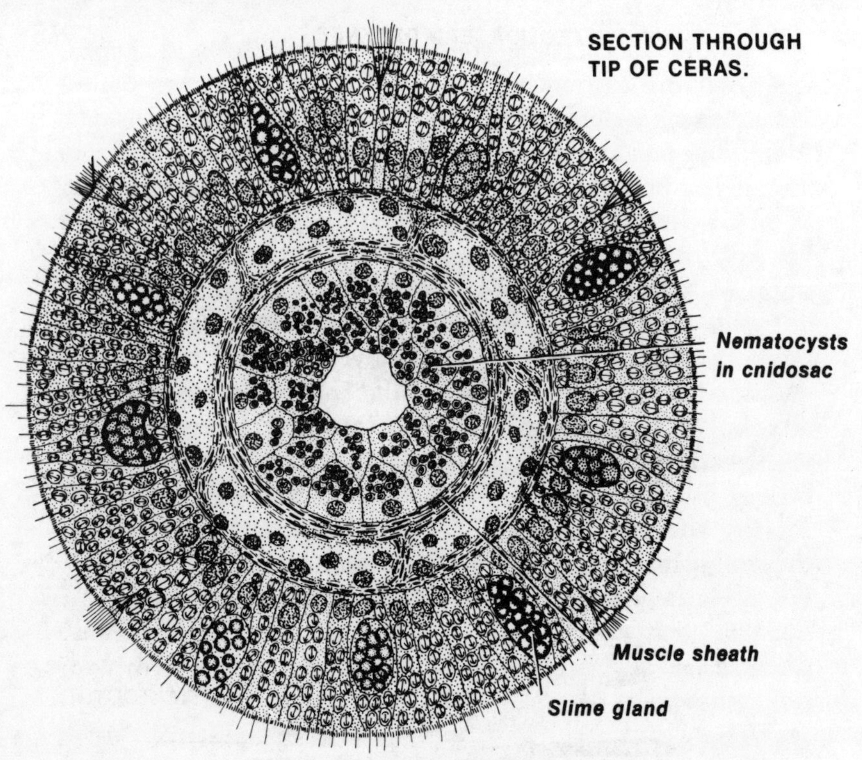

Physalia, and, moreover, only the largest (and probably the most effective and virulent) stinging cells of that prey organism. Other types of nematocysts are observed to be broken down in the digestive gland. Glaucid nudibranchs are, in fact, somewhat dangerous for human beings to handle without protection, and painful stingings by *Glaucus* have been reported by bathers in Australia. The Caribbean aeolidacean *Learchis poica* feeds on the "fire hydroid," *Pennaria distichia*, (whose stinging cells can raise ugly and painful wounds on the human skin) and is potentially dangerous in a similar way to the Indo-Pacific glaucid nudibranchs.

The utilization of coelenterate nematocysts for defense occurs throughout the aeolidaceans, with the exception of *Calma glaucoides* and various species of *Favorinus*, which do not habitually feed on Cnidaria, and of *Phestilla* species, which feed on living corals (*P. melanobrachia*, for example, attacks the ahermatype corals *Dendrophyllia* and *Tubastraea*) having extremely small nematocysts that are useless for molluscan defense. A similar defensive mechanism is found independently in the dendronotacean family Hancockiidae, but here the cnidosacs are found not only in the cerata but also beneath the skin of the flanks of the body; it is not yet known whether they are derived from the cnidarians upon which the nudibranchs are usually found.

When attacked, an aeolidacean nudibranch usually holds the cerata out erect and may even muster them together so as to aim them toward the enemy. In other words, attacks are invited toward the cerata and away from the vital head and visceral mass. Of course, such defensive adaptations by no means render their possessors immune to all attacks; *Aeolidia papillosa*, for example, has been found in the stomachs of young haddock, *Melanogrammus aeglefinus*. But no one who has carried out behavioral tests in marine aquaria could doubt that the principal method of defense possessed by aeolidaceans is the emission of clouds of exploding nematocysts after an attack by fishes.

The doridacean nudibranchs often possess calcareous spicules in the skin which may be dagger-shaped and aggregated into prickly papillae on the upper surface of the mantle. Sometimes, as in the Pacific American *Laila cockerelli*, these may take the form of elongated finger-like papillae with brilliantly colored tips, rather like the cerata of aeolidaceans. It is possible that such a spiculose texture would diminish the attractiveness of dorids as food for fishes, and it is interesting that spiculose nudibranchs are refused as food by fishes in experiments.

In all the other nudibranchs that have received careful study, skin glands have been found whose position and function can only be explained satisfactorily as defensive. They are always present in addition to the usual mucous glands associated with ciliated epithelia. They are often present in greatest abundance in the areas of the skin which would be first encountered by an inquisitive predator in nature. In species which possess dorsal papillae projecting some way out from the body, these glands are usually concentrated in the papillae. Papillae of this kind are usually non-retractile and, like aeolidacean cerata, may be rapidly regenerated if damaged. On the other hand, those dorsal processes which do not contain these glands (such as the head tentacles or the gills of some nudibranchs) are always retractile and are quickly concealed if the animal is disturbed. In short, the behavior of these nudibranchs if molested roughly is always such as to bring about the concealment of those regions of the body which are not endowed with these glands.

The chemical make-up of the majority of these defensive secretions of nudibranchs is almost completely unknown, but it seems likely that they will prove to be proteins. Some of the secretions taste extremely unpleasant to the human tongue and often possess strong toxicity to marine test organisms. The defensive slime secreted by the Pacific nudibranch *Phyllidia varicosa* contains materials rapidly poisonous to various fish and crustaceans tested. This slime, like the secretions of many other nudibranchs, has a lingering and distinctive odor. Some nudibranch secretions, such as those of the Japanese *Kaloplocamus* and *Plocamopherus*, are strongly luminescent.

Many arguments have taken place about the significance of the coloration of nudibranchs which in so many cases appears to be meaningful. The cerata of aeolidaceans and the dorsal papillae of many dorids or of dendronotaceans are frequently patterned in such flamboyant ways as to foster the theory that their coloration serves in nature to attract the attention of a predator away from the less conspicuous (but more fragile) head and visceral mass. Some other nudibranchs, however, are extremely well camouflaged so as to elude visually searching predators. Unfortunately, we have at present very little accurate information about the method of selection of food by predatory fish in different kinds of marine habitats. One thing that is certain is that in the nudibranchs the usual dividing line between brightly colored, distasteful forms on the one hand, and cryptically marked, readily edible forms on the other, does not hold. *All* the nudibranchs that have been studied in aquarium tests proved to be distasteful to fish.

Larval Biology

The spawn-mass of a nudibranch may be globular or ribbon-like and of many shapes, sizes, and colors in the various species. The dorid nudibranchs produce spawn which takes the form of a ribbon attached along one edge, while many aeolidaceans and dendronotaceans lay spawn-masses which are cylindrical, capsule-filled cords attached along one side by a thin capsule-free sheet. In some very small aeolids the spawn takes the form of a small kidney-shaped jelly bag attached by one side.

In an Antarctic nudibranch, *Austrodoris macmurdensis*, the spawn is strongly reinforced by chitin-like and other proteinaceous materials as an adaptation for an unusually long embryonic period (more than 50 days).

A general consideration of the range of shape and size of nudibranch spawn-masses leads to the following conclusions:

There is a positive correlation in nudibranchs between egg size and the length of the embryonic period and an inverse correlation between both and batch size.

The largest species of nudibranch adults tend to produce larger eggs in greater numbers.

Within any species, the larger individuals tend to produce more eggs than do the smaller individuals.

Within any species, egg size varies little through the geographical range, although measurable effects of salinity upon egg size, embryonic period, and fecundity have been reported for the North Atlantic *Embletonia pallida.*

An individual nudibranch produces its largest egg masses early in the breeding season; later masses contain smaller numbers of eggs.

The veliger larvae hatch after a total embryonic period that varies from 5-50 days in different species. Unlike prosobranch gastropods, the nudibranchs are true hermaphro-

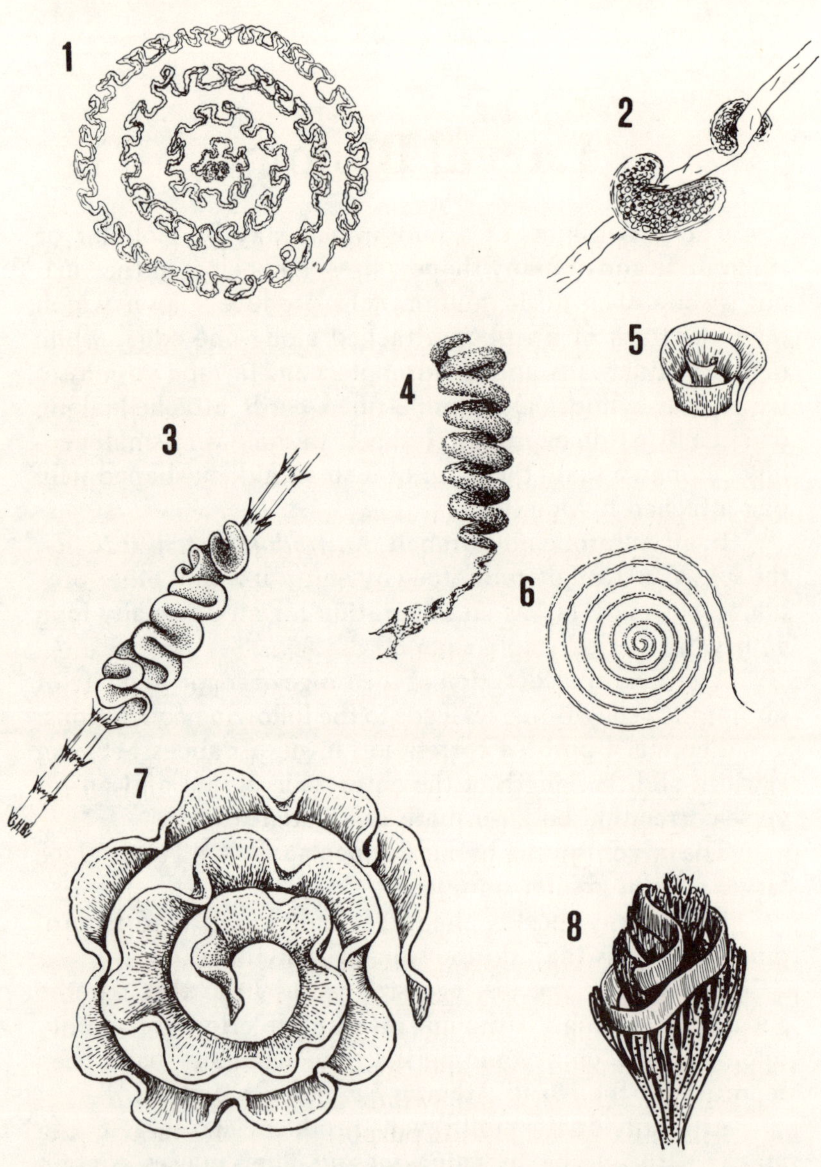

Spawn of nudibranchs. 1: *Aeolidia papillosa*; 2: *Catriona aurantia*; 3: *Doto fragilis*; 4: *Cumanotus beaumonti* (after Hurst); 5: *Ancula cristata*; 6: *Favorinus branchialis*; 7: *Acanthodoris pilosa*; 8: *Greilada elegans* (on bryozoan).

dites; each individual has both an ovary and testis and functions as both male and female (often simultaneously), so that fecundity is high. Some larger nudibranchs may produce up to a million larvae.

Details of larval behavior and ecology have been established for a few species of nudibranchs. Perhaps the best known of these is the dorid *Adalaria proxima*, and the following account is based upon this species.

The veligers of *Adalaria*, after escaping from the egg mass, swim strongly upward. In laboratory culture this almost invariably results in their becoming trapped in the surface film of the water, for the larval shell has a peculiar water-repellent quality. The larvae become lodged in the meniscus with the back of the shell projecting out of the water; the velum and foot remain beneath the surface, so the ciliary apparatus of the larva is unaffected. In order to allow further development to take place it is necessary to sink such marooned larvae by gently squirting water from a pipette onto the surface.

Larval locomotion is effected by the beat of the long velar cilia, which imparts a forward motion to the larva. The velum is well developed in *Adalaria* and the locomotor cilia are long in proportion. The larva can swim in a straight line, but the larvae of most other nudibranchs spiral as they proceed. Swimming activity is interrupted at intervals, when the larva half retracts into the shell and sinks slowly. If at any time while swimming the larva encounters an obstacle, it creeps over it in the usual gastropod fashion, propelled by the cilia of the foot. During creeping the velar lobes are brought together and held partially within the shell. If the obstacle is not suitable for metamorphosis of the larva, the velar lobes are protruded and the larva swims away.

Feeding occurs during swimming, for the beat of the long velar cilia serves a dual purpose for both imparting a forward motion to the larva and in bringing a constantly renewed supply of sea water within reach of the cephalo-pedal ciliary apparatus. Particles in the water are then conducted swiftly toward and into the mouth by the cilia on the subvelar ridges. In *Adalaria* feeding is of relatively small im-

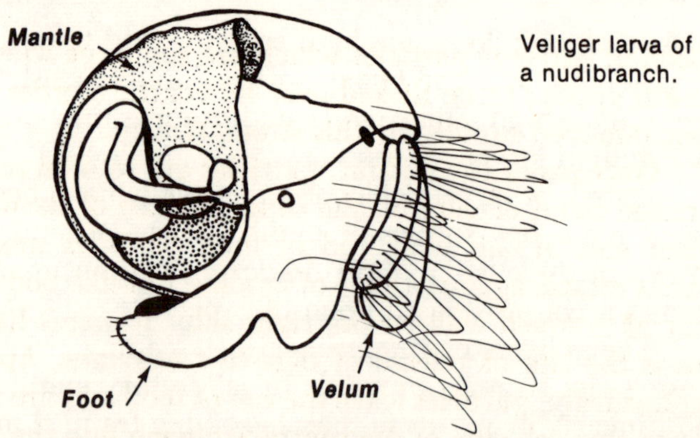

portance and normal further development can occur in larvae reared in sterile, filtered sea water. In most other nudibranch larvae, however, feeding is essential for the promotion of progressive morphogenetic changes.

The borders of the mouth are strongly ciliated, and particles of sufficiently small size (0.010-0.015 millimeters in diameter in *Coryphella lineata* and in *Archidoris pseudoargus*) are impelled into the foregut. The external surface of the foot is ciliated, these cilia being particularly strongly developed along a narrow zone leading in many species of nudibranchs from the ventral border of the mouth to the papillose tip of the foot. This narrow band forms the main

Larval metamorphosis in *Adalaria proxima*. 1: Aggregation and blending of the major ganglia. 2: Reflexion of the mantle. 3: Loss of the shell. 4: Loss of the larval retractor muscle. 5: Loss of the operculum. 6: Loss of the larval kidney. 7: Enlargement of the digestive gland.

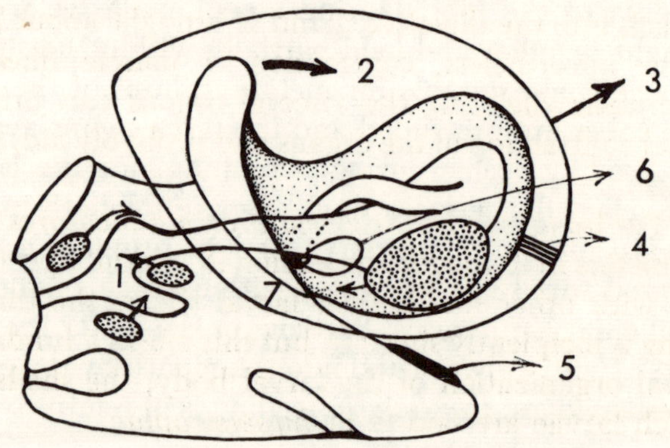

rejection current; particles which are too large to enter the mouth appear to be flicked over its ventral border and are then removed rapidly by this strong stream.

Once inside the midgut, particles are rotated rapidly by a raised band of cilia on the inner surface of its wall. The lower end of this loose rod of food particles mixed with mucus rotates against a zone of hyaline rod-like bodies in the ventral stomach wall on the right side. It seems likely that this is the site of release of digestive enzymes. Apparently haphazardly, particles leave the rod of food and are impelled by the sluggish cilia of the ventral stomach into the left midgut diverticulum or digestive gland. (The right midgut diverticulum functions for yolk storage and plays no part in larval

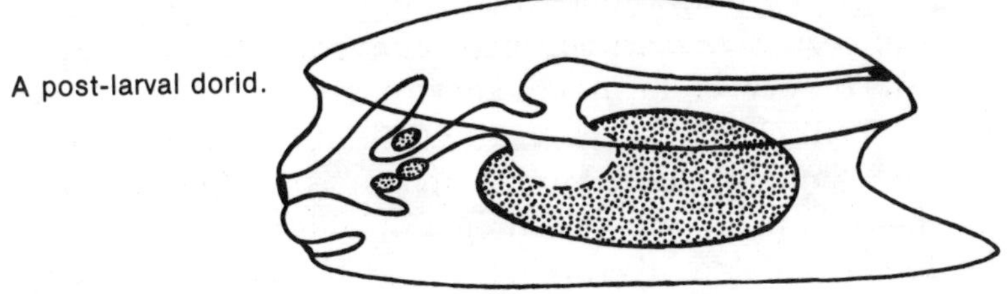

feeding.) The digestive gland in nudibranch larvae is usually pale greenish brown in color and has a wall of single-cell thickness. Sections through digestive glands of larvae of, for instance, *Adalaria proxima* or *Archidoris pseudoargus* that have been allowed access to unicellular algae like *Phaeodactylum tricornutum* or *Isochrysis galbana* show intracellular food vacuoles to be present. The passage of particles from the stomach into the digestive gland is brought about by ciliary means; nevertheless, contraction of muscle fibers in the membranes enclosing the visceral organs can bring about partial contraction of the organs and probably aids this interchange.

The larval shell of *Adalaria proxima* measures 0.28-0.30 millimeters in its longest dimension in lateral aspect. In common with other nudibranch larval shells, the direction of coiling is incipiently sinistral, but this masks a fundamentally dextral organization of the larval body; the shells of nudibranch larvae are said to be *hyperstrophic*.

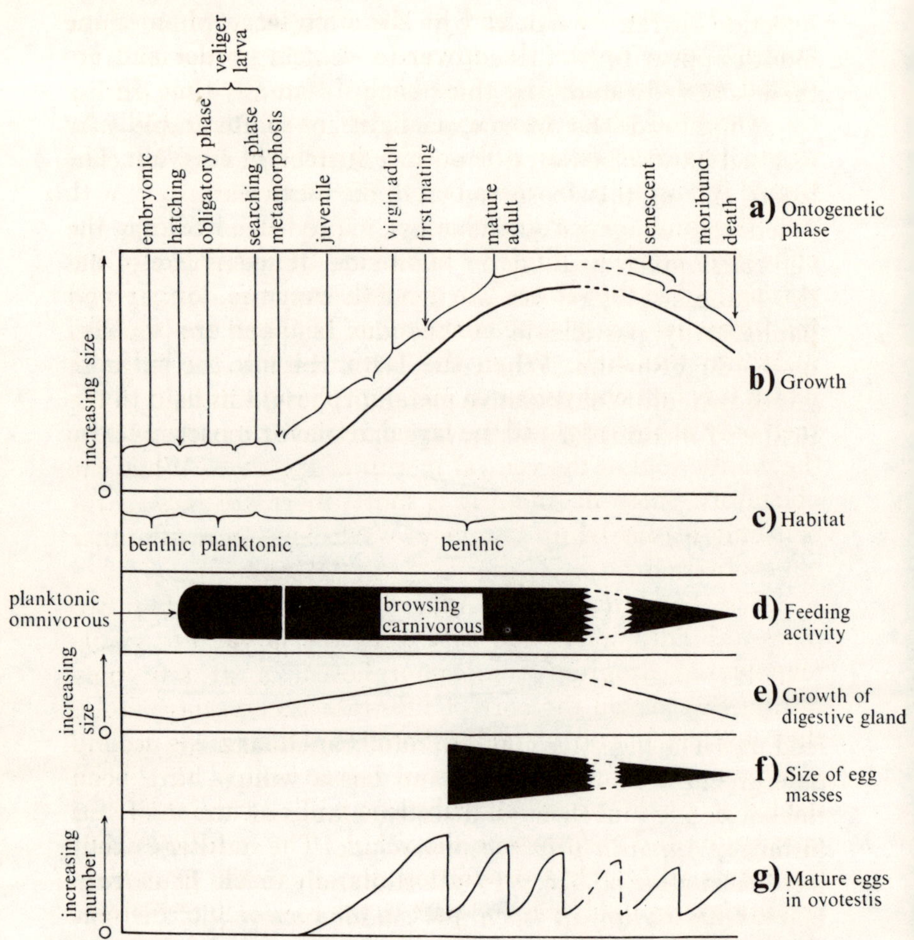

Essential features of nudibranch life cycles.

The larval life of *Adalaria proxima* may be divided into two distinct stages, the first occupying one or two days after hatching, the second of up to 14 days' duration. During the first stage the larvae are negatively geotactic (repulsed by a substrate); this behavior is reversed at the start of the second stage. It is convenient to name the first stage the *obligatory* phase of pelagic life, for until the internal changes which take place during it are completed the larva is incapable of metamorphosis. At the end of the obligatory phase these changes are complete, the upward-swimming behavior is reversed, and the *searching* phase begins. If at any time during the searching phase the larva alights on a favorable substrate, metamorphosis will occur. Searching may be continued for up to two weeks before the larva dies.

The most important changes which occur during the obligatory phase in *Adalaria* larvae are the reflection of the mantle fold so that it lies like a saddle over the dorsum and the severance of several connections between the visceral mass and the shell. When the larva reaches the stage at which it is first competent to metamorphose, it is held to the shell only at the origin of the larval retractor muscle and by the mantle fold at the ventral mouth of the shell. While the obligatory phase in *Adalaria* is short, there are reasons for believing it may be up to several weeks long in many other opisthobranchs.

The searching phase in *Adalaria* can be divided into two parts: (a) Throughout the first week the larvae are viable (capable of progressive metamorphosis); if at any time during this period the correct substrate becomes available, they will settle and metamorphose. (b) During the second week an increasing proportion of larvae which have been denied access to the correct substrate will cast the shell and operculum and even reduce the velum. The resultant shell-less forms, although a proportion remain viable in culture conditions, probably play no part in the economy of the species in nature.

Settlement and metamorphosis in *Adalaria* larvae will occur only on a live colony of the bryozoan *Electra pilosa*, which is the preferred prey of the adults. Settlement will not occur on other species of shallow water bryozoans employed in laboratory tests. Larval discrimination appears to be chiefly chemosensory. It has been shown to be remarkably precise, not only in *Adalaria* but also in *Tritonia hombergi*, *Eubranchus exiguus*, and *Phestilla sibogae*. It is clear that such behavioral mechanisms must bring about in nature the establishment of populations of these species in areas favorable for postlarval development.

The life cycles of surprisingly few species of nudibranchs have been worked out in detail. In those North Atlantic species that have received close study, the maximum life span has proved to be one year or less. Some of these species have annual life cycles, with one breeding period per annum (but occasionally two as in *Onchidoris bilamellata*), while others may pass through numerous generations in a year. The purely annual species tend to feed on organisms which have one conspicuous quality in common: their extreme abundance and stability in certain types of localities at all seasons of the year.

Those nudibranchs which are known to pass through a number of generations in a year are all species that feed upon more or less transitory prey, such as hydroids, which may spring up seasonally on submarine surfaces and be cropped to extinction within a few days. Nudibranchs which attack such organisms possess certain attributes in common. They often grow to sexual maturity very rapidly, reducing the risk of dying without progeny; they are active and voracious, and many of them are known to attack a wide variety of food-organisms; juveniles and adults are often found side by side, and spawning tends to occur through much of the year. A distinct lack of synchrony of the life cycles of the individuals composing the population is characteristic of these hydroid-eating nudibranchs, contrary to the orderly, rather synchronous sexual development and spawning behavior of the sponge-, soft coral-, and bryozoan-feeding nudibranchs, many of which are tied to a single prey organism all through the benthic phases of the life cycle.

There is a great deal of evidence that new populations of nudibranchs are commonly established in favorable regions by the exercise of the ability of the searching veliger to recognize and metamorphose upon some component of the diet of the benthic stages. That they possess this ability is proven in the case of many species and indicated by circumstantial evidence in many others. What is not known is whether the larvae of those nudibranchs which as adults feed upon a variety of organisms have less precisely specific settlement preferences than those of species with a more rigid adult diet.

Zoogeography

This is a fascinating subject and one that is in its infancy so far as the distribution of nudibranchs is concerned. Everyone knows that nudibranchs (or, for that matter, any other species or group of animals or plants) are not to be found randomly over the surface of the Earth. Each species or group of species appears to have a preferred geographical range. In some nudibranchs this range may be very wide; *Glaucus atlanticus*, for example, occurs in the surface waters of the three major oceans, and *Aeolidia papillosa* may be found in both northern European and Californian waters. But these ranges are less wide than might at first appear: there are restrictions. *Glaucus atlanticus* occurs principally within the tropics, and *Aeolidia papillosa* is to be found chiefly in cool temperate waters.

Some species of nudibranchs pose zoogeographical puzzles. *Chromodoris loringi* and *C. marginata* are two small dorid nudibranchs which are abundant in bays near Sydney in New South Wales, Australia, but have up to the present time never been recorded elsewhere. *Notodoris gardineri* shows us another perplexing kind of distribution; it has been recorded only very rarely during the last century (although it is a large and conspicuously marked species), but these records have come from widely scattered localities over an enormous oceanic area embracing both Pacific and Indian Oceans. To a certain extent these puzzles arise from our sparse knowledge, and it may be confidently predicted that continued accurate systematic research work will resolve many of our present problems. But at the moment we can only work with the available imperfect distributional data and try to make some guesses about the causal interrelationships between nudibranch distribution and nudibranch physiological ecology.

We are on safe ground when we argue that if a nudibranch species (let us call it *Doris X*) occurs commonly in a certain marine locality or sea area, then conditions there must be favorable for *Doris X* to grow and to flourish. Furthermore, if we have records which prove that *Doris X* has occurred in that area for some considerable time (over many years), then we can feel confident that conditions there must also be favorable for breeding and population renewal.

But we are on shaky ground if we fall into the philosophical trap of making the converse deduction, namely that the absence of *Doris X* from other marine localities proves that conditions there must be unfavorable for its survival. To begin with, this deduction rests upon negative evidence: the apparent absence of *Doris X* (and an eagle-eyed field collector may at any time render this obsolete with an unexpected find). In addition, it includes the unwarranted assumption that the patchwork of animal distribution is fixed and static, whereas everything we have discovered about organic evolution indicates that changes are still going on. Species are still becoming extinct and, conversely, new species continue to arise. The geographical ranges of animal and plant species are waxing and waning, although looking at the present-day fauna and flora we cannot accurately decide which species are increasing and which are declining. In short, when we note that *Doris X* is unrecorded for, let us say, San Francisco Bay, we cannot decide between the following three possibilities: (a) *Doris X* is there but has yet to be found; (b) *Doris X*, although prevalent in neighboring localities, has been prevented from entering and living in the Bay by some local combination of physiological and ecological limitations; (c) *Doris X* has zoogeographical boundaries which have not as yet impinged on that oceanic area (in the example quoted, the northwestern Pacific Ocean).

Let us examine more closely these three possibilities. With regard to possibility (a), only repeated searching by a number of collectors using a full range of modern gear over a period of years can dispel our suspicion about the value of negative evidence in zoogeography. Possibility (b) is certainly a reality in many well documented examples. We know of

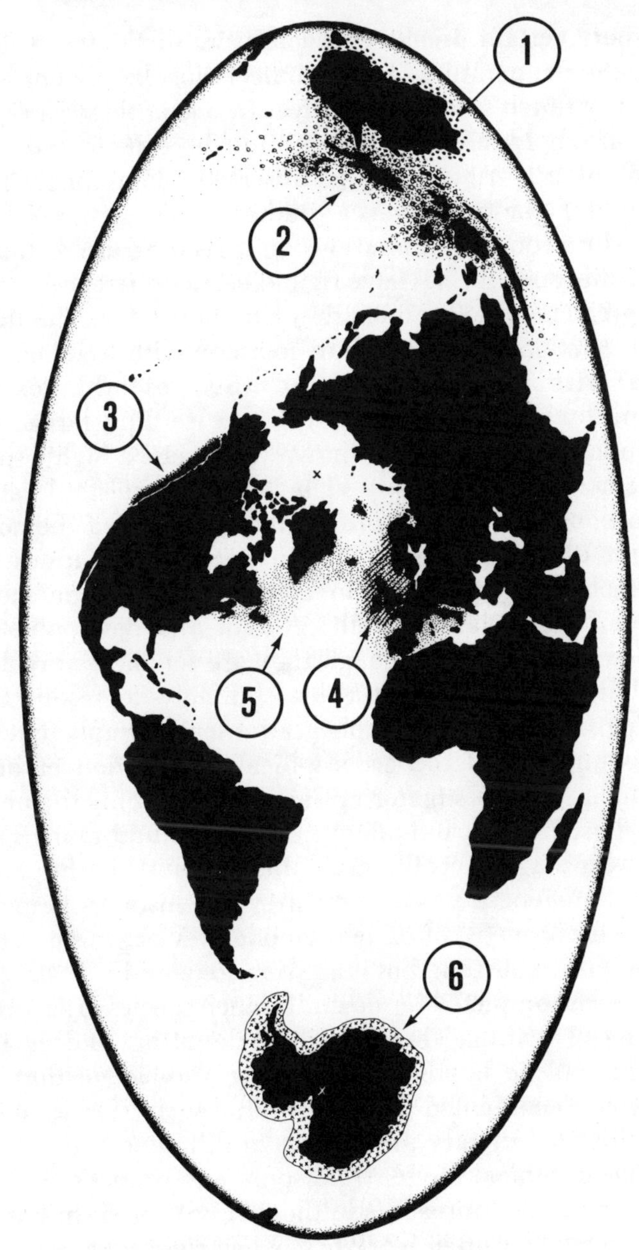

Some centers of distribution of nudibranch taxa. 1: Asteronotidae; 2: Chromodorididae; 3: *Cadlina* and *Dendronotus*;
4: Pseudodorids; 5: Aeolidacea; 6: *Austrodoris*.

cases where certain precise characteristics of the sea water (such as the temperature or the salinity) must be present before a nudibranch species can thrive. In one of these cases, a specific threshold temperature rise must be experienced before spawning is triggered, and this factor is probably responsible for limiting such a nudibranch's geographical spread. More common limitations on spreading are dietary. Many nudibranch species have rigid dietary preferences, and if their preferred prey is absent they cannot survive. This life-or-death relationship has much in common with well known host-parasite relationships, and this resemblance is heightened when one discovers that the settling larvae of many nudibranchs have been shown to exercise a highly specific chemosensory selectivity such that they will settle and metamorphose only on the principal component of the normal adult diet. This means that in nature populations of nudibranchs are established in areas which will be most favorable for post-larval (adult) growth and development.

These facts, exciting though they are for the embryologist and for the larval biologist, are chillingly depressing for the ecologist or the zoogeographer because they imply that in order to understand the geographical distribution of any nudibranch, the investigator must analyze not only the physiological preferences and tolerances of that nudibranch species at every stage in its life cycle, but he must do the same for every ontogenetic phase of the prey organism or, in more catholic nudibranchs, of more than one prey organism. This is not an impossible task, but it is a very long and tedious one and has been completed for no nudibranch species or its prey at the time of writing. The task will be daunting, and the investigator will be handicapped by the knowledge that in some well documented cases a nudibranch species may exhibit different dietary preferences in different parts of its known geographical range. (And some nudibranchs change their diet as they approach adulthood.) But an enthusiastic research team can often work best when the obstacles seem most difficult to surmount, and it can only be a question of time before someone succeeds in piecing together the physiological and ecological balance sheet that will for the first time enable us to understand how the physical and biological factors that govern life processes act together so as to control and regulate zoogeographical distribution of a nudibranch.

Now we must turn to a consideration of possibility (c), which can be re-stated thus: is there any evidence from the zoogeography of nudibranchs that the spread of the boundaries of certain forms is continuing? To crystallize this question into an actual, not a hypothetical, framework we can consider the case of the North Pacific nudibranch *Tochuina tetraquetra*. This is prevalent from Japan to Alaska, and its North American distribution extends southward down the coast of California as far as Monterey Bay. Given time, will it one day penetrate further south? Should we expect to encounter publications in the future recording *Tochuina* from, for instance, Baja California, then later from Acapulco? Is time all that is needed for the further spread of *Tochuina*?

Unfortunately, it is a lot easier to ask such apparently simple questions than to answer them. But we do have some evidence, albeit indirect, about the probably dynamic nature of some nudibranch families and their oceanic distribution. This evidence comes from the study of suprageneric (above the level of the genus) systematics in the nudibranchs. If we can find evidence of the preponderance of certain great natural groups of nudibranch species in significantly circumscribed areas, then this could plausibly fit into the general proposition that nudibranch zoogeographical boundaries are by no means frozen. And if we can identify such boundaries, we might well be able to predict future large-scale spreads of certain kinds of nudibranchs. First of all, what kind of evidence do we have?

The coasts of North America are especially rich in materials of the kind needed to further pursue our enquiries. If we travel to the Pacific Northwest to the state of Washington and then embark on the ferry to the San Juan Islands, we come to the township of Friday Harbor, on San Juan Island itself. From the famous marine research laboratories at Friday Harbor it is possible to capture, using aqualung equipment, large numbers of shallow-water nudibranchs. Among the catch may be numerous species of the delicate nudibranch genus *Dendronotus*: on a good day we may find *Dendronotus frondosus*, *D. iris* (= *giganteus*), *D. robustus*, *D. dalli*, *D. rufus*, *D. gracilis*, *D. albus*, *D. subramosus*, *D. diversicolor*, and *D. albopunctatus*. The truly remarkable feature of this set of finds is that it contains about 80% of the world's species of *Dendronotus*. Nowhere else is it possible to find more than one or two species of *Dendronotus*, and these are apparently all from the Northern Hemisphere. Another example from the same area is afforded by the doridacean genus *Cadlina*, which in the vicinity of the San Juan Islands can yield on a good day's collecting *Cadlina flavomaculata*, *C. luteomarginata*, and *C. modesta* with, further north, *C. pacifica* and *C. laevis*. This is almost the total world tally for species of *Cadlina*, although one of these species, *C. laevis*, has a wide range and is known to occur also from the North Atlantic across to the coasts of Europe.

Before we start to consider the significance of such aggregations, let us look at one more well documented example, the case of the Onchidorididae, containing the genera *Onchidoris*, *Adalaria*, and *Acanthodoris*. These are sometimes termed "pseudodorids" because, although they look very much like common large doridacean nudibranchs such as *Archidoris* and *Doris*, they have a narrow radula, no branchial pocket, and, in short, are not especially closely related to the "true" dorids. The peculiarity of their distribution is that it is in the cool North Atlantic (although a very few species have extended further), centered on or near northern Europe. From the British Isles alone there have been descriptions of *Onchidoris bilamellata*, *O. muricata*, *O. luteocincta*, *O. depressa*, *O. inconspicua*, *O. sparsa*, *O. oblonga*, *O. pusilla*, *Acanthodoris pilosa*, *Adalaria loveni*, and *A. proxima*. This total far exceeds the number of species of Onchidorididae recorded from the rest of the world.

These are three well documented examples of a phenomenon which has countless other examples. The Chromodorididae predominate in the Pacific Ocean, the aeolidaceans in general appear to be centered on the Mediterranean Sea, *Aphelodoris* species are abundant only in Australasia, and *Austrodoris* species seem to be tied to the very coldest waters in the southern oceans. Every nudibranch specialist will have his own favorite example! But what does it all mean?

It seems that in these examples we are seeing 'cradles' for the evolution of certain successful groups of nudibranchs. As an academic zoologist might say, we are able to identify probable centers of evolutionary adaptive radiation. As more new species are discovered (the author's personal estimate is that only about 50-60% of the total living nudibranch species have as yet been recorded by scientists) and as more of the misidentifications of the past are corrected and cleared up (very many errors remain to be eradicated), it appears likely that more such centers of adaptive radiation of nudibranchs will be pin-pointed.

But a note of caution. The oceans have not always been as they are now. In the Cretaceous period 80 million years ago the land masses and the oceans looked very unlike their present sizes and positions. The fossil record tells us that the area of greatest coral diversity and luxuriance was then in the area which now constitutes our Mediterranean Sea. That was as rich in corals in the Cretaceous as the Malaysian archipelago is today. If we tried to argue that the center of reef coral evolutionary adaptive radiation was the Malaysian archipelago simply because that is where the greatest coral diversity is encountered today, then we would be making a bad mistake. The lesson to be learned is that we must be wary in our interpretation of nudibranch systematic ecology today because we have no fossil history to check our theories against.

This is, however, a very exciting time in the development of the subject. Every find, every SCUBA dive, every visit to a fresh locality, and every accurate description of a live collection of nudibranchs will yield the data so much needed if we are to make sense of some of these trends in nudibranch zoogeography.

Glossary

ARBORESCENT — Much-branched, tree-like.

BLASTOCOELIC CAVITIES — Splits between cells or groups of cells which appear during early development in many animals and which in the mollusks persist into the adult phase.

BUCCAL APPARATUS — The rasping tongue (radula) of a mollusk and its associated musculature, with the frequent addition of a pair of stout horny jaws.

COELOM — Body cavity lined by mesodermal tissue, characteristic of higher animals (i.e., above coelenterates, flatworms and nematodes).

EPIFAUNA — Animals that dwell upon the ocean bottom (including the shore).

FOLIACEOUS — Bearing many leaf-like appendages.

GANGLIA — Bulbous aggregations of nerve cell nuclei with linking fibers.

HYPERSTROPHIC SHELL — This is found in those few mollusks in which the shell coils in a certain direction whereas the soft parts within it coil in the opposite direction. Nudibranch larvae are hyperstrophic.

MANTLE CAVITY — Cavity housing the gills in typical mollusks, becoming more open in advanced forms; completely lost in the highest nudibranch mollusks.

OPISTHOBRANCH — Mollusks having evolved from prosobranch ancestors and possessing the following combination of characters: reduced shell and operculum, hermaphrodite genital apparatus, mantle cavity organs shifted toward the rear of the body.

PALLIAL — (Adjective) Relating to the molluscan mantle.

PROPODIAL — (Adjective) Relating to the front part of the molluscan foot.

PROSOBRANCH — Mollusks which retain the primitive features of ancestral gastropods, namely a well developed shell and operculum, separate sexes, and mantle cavity organs situated near the front of the body.

PULMONATE — Gastropod mollusks which have evolved in such ways (tough skins and the ability to breathe air through a lung are commonly encountered) that they can live successfully on land. Common pulmonates are the garden slugs and snails.

SACOGLOSSANS — Gastropod mollusks resembling nudibranchs but distinguished from them by their adaptations for a herbivorous diet; nudibranchs are all carnivorous.

SPICULAR CONCRETIONS — Calcareous (limey) or siliceous spines, disks, spheres, or asters strengthening the skin of an animal.

TORSION — An evolutionary process unique to the gastropod mollusks; it has resulted in a twisting between the head/foot and the visceral mass, so that the originally posterior mantle cavity comes, in the Streptoneura (or Prosobranchia), to lie anteriorly just behind the head.

Recommended Reading

Abe, T. 1964. *Opisthobranchia of Toyama Bay and adjacent waters.* Tokyo, Hokoryu-Kan. Pp. 95.

Baba, K. 1949. *Opisthobranchia of Sagami Bay collected by His Majesty, the Emperor of Japan.* Japan, Iwanami Shoten. Pp. 194.

Morton, J.E. 1958. *Molluscs.* London, Hutchinson. Pp. 244.

MacFarland, F.M. 1966. *Studies of Opisthobranchiate Mollusks of the Pacific Coast of North America. Mem. Calif. Acad. Sci.* Volume 6. Pp. 546.

Thompson, T.E. 1976. *Biology of Opisthobranch Molluscs. Volume I.* London, The Ray Society. Pp. 206.

Yonge, C.M. and Thompson, T.E. 1976. *Living Marine Molluscs.* London, Collins.

Illustrations Index

Page numbers in **bold** type refer to color photos.